Geology Terms

in English and Spanish

Terminología Geológica
en Español e Inglés

Edited by
Barbara Belding Birnbaum
and Francisco Súarez

Sunbelt Pocket Guide
A series edited by Lowell Lindsay

Sunbelt Publications
San Diego, California

Geology Terms in English and Spanish
Copyright © 2000 by Sunbelt Publications, Inc.
All rights reserved. First edition 2000, third printing 2011.

Edited by Barbara B. Birnbaum, San Diego Association of Geologists (SDA◆
and Francisco Suárez, CICESE
Cover and book design by Fred Noce
Project coordination by Joni Harlan and Debi Young
Composition by Lynne Bush
Printed in the United States of America by 360 Digital Books

Sunbelt Publications, Inc.
P.O. Box 191126
San Diego, CA 92159-1126
(619) 258-4911, fax: (619) 258-4916
www.sunbeltbooks.com

"Sunbelt Pocket Guides"
A Series edited by Lowell Lindsay

15 14 13 12 11   5 4 3

Library of Congress Cataloging-in-Publication Data
Aurand, Henry, 1924-
  Geology Terms in English and Spanish = Terminología geológia en español e ing◆
Henry Aurand. – 1 st ed.
  p. cm. — (Sunbelt Pocket Guide)

  Includes bibliographical references.
  ISBN 978-0-932653-29-1

  1. Geology Dictionaries. 2. English language Dictionaries—Spanish. 3. Geology
Dictionaries—Spanish. 4 Spanish language Dictionaries—English. I. Title. II. Title
Terminología geológia en español e inglés III. Series.

QE5.A88 1999   99-19537
550'.3—dc21   CIP

Front cover photo: Photodisc, copyright 1999

Contents
Indice

Preface   v

Prefacio   vii

List of Charts and Tables *Lista de Cartas y Tablas*   ix

Abbreviations *Abreviaciones*   x

Spanish Pronunciations   xi

English-Spanish Geology Terms   1

Español-Inglés Terminología Geológica   51

Charts and Tables *Cartas y Tablas*   105

References *Referencias*   116

Preface

My purpose in writing this book was to help those who are reading, writing, or translating geologic material in either Spanish or English. My overriding goal was to make the book useful in a wide range of settings while keeping it compact and affordable. Successful abridgement is an arduous and difficult process and brings to mind the variously attributed quotation "I am sorry to have written you such a long letter, but I did not have the time to write a short one." To facilitate abridgement I used the following guidelines when deciding whether to include a particular word or term:

- Include frequently used terms.

- Give precedence to Mexican Spanish usage and then Latin American. As with all global languages, there are different words and terms to describe the same thing. Many of the terms contained herein may vary in accuracy and applicability, depending on the country or circumstances.

- Include some terms from the related specialties of mineralogy, paleontology, and seismology and some fewer from hydrology, meteorology, volcanology, and geomagnetism. For mineralogy, include the names of the most common minerals, particularly those that are primary constituents of common rocks. For paleontology, include the names of animal fossils only to the phylum level. Additionally, include descriptive terms (colors, sizes, and shapes) commonly used in fieldwork.

- Include synonyms only if they are in common use.
- Limit the overall number of words and terms in an effort to meet the primary objectives of convenience and affordability.

For Spanish words, gender and parts of speech have been indicated only in cases where there might be any doubt or difference from regular forms. For instance, the abbreviation *f* follows the word *corriente,* indicating that the word is feminine, whereas no gender designation follows the word *caliza,* because, in this case, its spelling indicates the gender (a list of abbreviations follows the preface).

I wish to thank the publisher, Lowell Lindsay; the designer, Fred Noce; the production coordinator, Joni Harlan; the compositor, Lynne Bush; the production assistant, Jennifer Lindsay; and the editors, Barbara Belding Birnbaum and Francisco Súarez, for all their efforts, help, and advice.

I hope this book meets your needs. I encourage you to forward comments, suggestions, and corrections to me in care of the publisher, Sunbelt Publications.

Henry Aurand
Solana Beach, California
October 1999

Prefacio

La intención de este libro es ayudar los que leen, escriben o traducen materias geológicas, en inglés e español. Mi meta última es ensurar que este libro sea conveniente para situaciones varioses, y también mantener al tamaño y el valor del producto.

Por eso, abreviar el libro es necesario. Hay un adagio inglés que dice, "Discúlpame para escribiéndose una carta tan larga; no tenía tiempo para escribir una corta." Ésto se aplica a este libro. Abreviarlo con éxito es una tarea muy difícil y dura. En este proceso, yo he usado los principios siguiendos en decidir si una palabra o un término debe ser incluido o no. No obstante le selección frecuentemente tiene que ser algo arbitraria. Entre los criterios usados en este libro fueron:

- Los términos de frecuente uso deben incluirse.

- El idioma usado en México debe tener precedencia. Con lenguajes mundiales, hay palabras y términos diferentes que describen la misma cosa. Muchos de ellos contenido en este libro pueden variar en aplicabilidad y en precisión, según el país o las circunstancias.

- Un pocos términos de las especialidades relacionados de mineralogía, paleontología, sismología deben ser incluidos, y menos de hidrología, meteorología, vulcanología, y geomagnetismo. De mineralogía fueron incluidos los nombres de minerales notables, especialmente los que son componentes primarios de rocas comúnes. De paleontología fueron incluidos los nombres de animales fósiles sólo al nivel de los filos. También fueron incluidos algunos términos descriptivos (colore, tamaño, forma, etc) que son usados frecuentemente an el campo.

- Los sinónimos deben ser incluidos solamente si son usados frecuentemente.
- El numero total de palabras y términos debe ser limitado para preservar los objetivos de conveniencia y precio razonable.

Para las palabras españoles, he indicado el género y las formas gramáticas solamente cuando hay una diferencia de las formas regulares. Por ejemplo, la *corriente* está identificada como feminina con una efe, pero la *caliza* no está identificada en cuanto a género. Una lista de abreviaciones sigue el prefacio.

Me gustaría dar gracias a los publicador, Lowell Lindsay; al diseñador, Fred Noce; a la coordinadora del producción, Joni Harlan; a la cajista, Lynne Bush; a la ayudante del producción, Jennifer Lindsay; y a los editores, Barbara Belding Birnbaum y Fransisco Súarez, para sus esfuerzos, ayudas, y consejos.

Espero que este libro satisfaga sus necesidades. Le animo a contactarnos por el publicador, Sunbelt Publications, con comentos, sugerencias, y correcciónes.

Henry Aurand
Solana Beach
Octubre 1999

Charts and Tables
Cartas y Tablas

Geologic Time Scale *Chronología Geológica*   106

Igneous Rock Chart *Diagrama de Rocas Igneas*   108

Grain Size Table *Tabla con Tamaños de Granos*   110

Prefixes for SI Unit Multiples *Prefijos de Multiples de Unidades SI*   112

Stratigraphic Abbreviations *Abreviaciónes Stratigráficos*   113

Conversion Chart of Weights and Measures *Tabla de Conversiones de Pesas y Medidas*   114

Abbreviations
Abreviaciones

adj	adjective	adjectivo
adv	adverb	adverbio
f	feminine	feminino
m	masculine	masculino
n	noun	nombre
p	plural	plural
s	singular	singular
v	verb	verbo

Spanish Pronunciations

Each Spanish vowel has one basic sound:

a is pronounced "ah"
e is pronounced "eh"
i is pronounced "ee"
o is pronounced "oh"
u is pronounced "oo"

Other letters in the Spanish alphabet with unique sounds:

b and *v* are pronounced as a soft *b*
j is pronounced as an *h*
ll is pronounced as a *y*
ñ is pronounced "ny"
z is pronounced as an *s*
rr is pronounced as a rolled *r*

To pluralize a Spanish noun, add *s* if the word ends in a vowel and *es* if it ends in a consonant.

English - Spanish

—A—

aa aa, lava viscosa
ablation ablación
abrade raer
abrasion raedura
abrasive abrasivo
absorb absorber
abutment lindero
abyss abismo, sima
abyssal abisal
abyssal current corriente
   abisal *f*
abyssal plain llanura abisal
accident accidente *m*
accretion acreción
achondrite acondrito
acid ácido *m*
acidic ácido *adj*
active volcano volcán activo,
   volcán en actividad *m*
adit galería de excavación
adobe adobe *m*
aerial aéro
aerobic aerobio
aerosol aerosol *m*
aftershock replica

agate ágata
age edad
aggrade elevar con
   sedimentación
aggregate agregado
air aire
alabaster alabastro
albite albita
alga alga
algae algas
algal algáceo
alidade alidada
alkali álcali *m*
alkaline alcalino
allochthonous alóctono
alloy aleación, liga
alluvial aluvial
alluvial fan abanico aluvial,
   cono aluvial
alluvial plain llanura aluvial
alluvium aluvión *m*
alpine alpino
alteration alteración
altered alterado
altimeter altímetro
altitude altitud *f*

3

alum alumbre *m*
alumina alumina
aluminum aluminio
amalgam amalgama
amber ambár *m*
amethyst amatista
ammonite amonite *mp*
amorphous amorfo
amphibole anfíbol *m*
amphibolite anfibolita
amygdaloid amigdaloide *m*
amygdule amígdala
anaerobic anaerobio
ancient antiguo
andesite andesita
angle ángulo
angle of dip ángulo de inclinación
angle of repose ángulo de reposo
angular angular, anguloso
anhydrite anhidrita
anhydrous anhidro
anisotropic anisótropo, anisotrópico
anisotropy anisotropía

annual anual
anomalous anómalo
anomaly anomalía
anorthite anortita
anorthoclase anortoclasa
anorthosite anortosita
antarctic antárctico
Antarctica Antártida
antecedent antecedente
anthracite antracita
anticlinal fold pliegue anticlinal *m*
anticline anticlinal *m*
antimonite antimonita
antimony antimonio
apatite apatita
apex ápice *m*
aphanitic rock roca afanítica
aquatic acuático
aqueous ácueo, acuoso
aquifer acuífero
aragonite aragonita
arc arco
arch arquear, arco
Archeozoic Arqueozoico
arctic ártico

4

arenaceous arenáceo
argillaceous arcil-loso, argiláceo
argilliferous arcillífero
argillite argilita
argon argón *m*
arid árido
arkose arcosa
arkosic arcósica
arroyo arroyo
arsenic arsénico
artesian artesiano
asbestos asbesto
ash ceniza
ash fall lluvia de cenizas
ash shower lluvia de cenizas
asphalt asfalto
asteroid asteroide *m*
asthenosphere astenosfera
asymmetric asimétrico
asymmetry asimetría
atmosphere atmósfera
atoll atolón *m*
attitude posición, postura
attrition desgaste *m*
aurora auroras

austral austral
autochthonous autóctono
avalanche avalancha, alud *f*
average promedio
axes ejes *mp*
axial axial
axis eje *m*
azimuth azimut *m*
azurite azurita

— B —

backbeach playa posterior
backshore costa posterior
badlands malpaís *m*
bajada bajada
ballast balasto, ripio
band banda, lista, capa, veta
banded bandeado
banding bandeando
bank banco, terraplén *m*, orilla, margen *m*
bar barra, banco en río, restinga
barchan barján
barite barita
barrier barrera

5

barrier beach playa en barrera
basalt basalto
basaltic basáltico
base base *f*, lecho
base level nivel de base, nivel básico
baseline línea de base
basement basamento
basic básico
basin cuenca, hoya
basin and range cuenca y cordillera
batholith batolito
bathymetric batrimétrico
bathymetry batimetría
bauxite bauxita
bay bahía
beach playa
beach gravel grava de playa
beach ridge cresta de playa
bearing orientación, rumbo, marcación
bed capa, lecho, yacimiento, cama
bedding estratificación

bedding plane plano de estratificación
bedrock roca de fondo, roca de lecho, roca firme, roca sólida, lecho de la roca
belemnites belemnites *mp*
below abajo
belt zona, cinturón
bench banco
benchmark mojón *m*, punto topográfico
bend recodo *m*
bentonite bentonita
berm berma, arcén *m*
berrylium berilio
beryl berilo
bevel chaflán *m*, biselar
bight seno, cala, calita, rada
binary binario
biochemical bioquímico
biogenesis biogénesis *f*
biogenic biogénico
biology biología
biotite biotita
birth nacimiento
bismuth bismuto

bituminous coal carbón bituminoso, hulla
bivalves bivalvos
black negro
black ore mena negra
black smoker fumarola negra
blister ampolla
blizzard ventisca, nevasca
block bloque *m*
blood sangre
blow soplar
blowhole bufadora
bluff risco, peñasco, farallón
blue azul
blunt despuntado, embotado
bog pantano, ciénega
boil hervir
boiling spring manantial hirviente
bolson bolsón *m*
bone hueso
borax bórax *m*
border margen *m*
borderland zona fronteriza
bore taladrar, perforar
boreal boreal

boron boro
boss figura de relieve, protuberancia
botanic botánico
botanist botanista
botany botánica
botryoid botrioide *m*
botryoidal botrioidal
bottom fondo, piso
boulder pedregón *m*, bloque *m*, canto rodado
boulder clay till *m*
brachiopods braquiópodos
brackish salobre
braided anastomosado
branch ramificar, rama, ramal *m*
branching fault falla ramificada
brass latón *m*
breach brecha
break quebrar, romper
breaker rompiente *m*
breaking rotura
breakwater rompeolas
breccia brecha

7

brecciated brechoso
bridge puente *m*
bright brillante
brightness brillo
brine salmuera
broad amplio
broken roto, quebrado
bronze bronce *m*
brook chorrillo, arroyuelo
brown moreno
bubble burbuja
bulge comba
bulged pando
bulk, in bulk en bruto, en volumen
burial entierro
bury enterrar, sepultar
butte relicto de erosión, pedestal *m*

— C —

calcareous calcáreo, calizo
calcareous ooze cieno calcáreo
calcify calcificar
calcite calcita

calcium calcio
caldera caldera
caliche caliche *m*, tosca
Cambrian Cámbrico
camera cámara
camp campamento
canyon cañón *m*
capable capaz
cape cabo, punto
capillary capilar *m* or *adj*
cap rock roca de cubierta, cubierta de roca, capuchón
carapace carapacho, caparazón
carbide carburo
carbon carbono
carbonaceous carbonoso
carbonate carbonato
carbon dioxide dióxido de carbono
carboniferous carbonífero
carnotite carnotita
cascade cascada
cassiterite casiterita
cast molde *m*
cataclastic cataclástico

8

cataract catarata
catastrophic catastrófico
catastrophism catastrofismo
cave cueva
cave-in derrumbamiento
cavern caverna
cavity cavidad
cement cemento *m*, cementar *v*
cementation cementación
Cenozoic Cenozoico
centigrade centígrado
center centro
chalcedony calcedonia
chalcopyrite calcopirita
chalk creta, tiza
chamber cámara
channel canal *m*
chart carta
chasm sima
chemical química
chemist químico
chemistry química
chert sílex *m*
chevron pliegue en «V»
chimney chimenea

chip pedacito
chlorate clorato
chloride cloruro
chlorine cloro
chlorite clorita
chondrite condrita
chondrule condrula
chronology cronología
chrysoberyl crisoberilio
cinder ceniza, escoria
cinder cone cono de ceniza
cistern cisterna
clast clasto
clastic clástico
clay arcilla, barro
clayey arcilloso
claystone roca arcillosa, arcillita
cliff acantilado, risco, despeñadero, precipicio
climate clima
clinker escoria de hulla
clinometer clinómetro
clod terrón *m*
close cercano, cerrar
closed cerrado

cloud nube *f*
coal carbón
coal tar alquitran de hulla *m*
coarse tosco, grueso
coarse-grained (phaneritic) fanerítico
coast costa
coastal costero
coastal plain llanura costera, planicie costanera
coast line línea costera
cobalt cobalto
cobble guijarro, guijón *m*
cohesion cohesión
cohesive cohesivo
coke coque *m*
collapse derrumbamiento, derumbarse, derrumbe *m*, hundirse, desplome *m*
collect recolectar
colloid coloide *m*
colloidal coloidal
colluvium coluvio
color color *m*
column columna

columnar columnario, columnar
columnar jointing diaclasa columnar
comet cometa *m*
compact compactar, compacto *adj*
compaction compactación
compass brújula, compas *m*
competence competencia
competent competente
complementary complementario
complex complejo
composite compuesto *m* or *adj*
composite volcano volcán compuesto
compound compuesto *m* or *adj*
compress comprimir
compression compresión
compressive compresivo
concave cóncavo
concentric concéntrico
conchoidal concoidal
concordant concordante

concretion concreción
concussion conmoción
condensate condensativo
condensation condensación
condense condensar
conduction conducción
conductive conductivo
conductivity conductibilidad
conduit conducto
cone cono
conformity concordancia, acordanza
conglomerate conglomerado
connate water agua fósil
conodont conodonte *m*
consequent consecuente
consolidated consolidado
consolidation consolidación
contact metamorphic rock roca metamórfica de contacto
continent continente *m*
continental drift deriva continental
continental margin margen continental *m*

continental platform plataforma continental
continental rise elevación continental
continental shelf plataforma continental
continental shield escudo continental
continental slope talud continental *m*
contour curva de nivel, contorno, contornear *v*
control controlar
convergence convergencia
convex convexo
coordinate coordenada
copper cobre *m*
coprolite coprolito
coral coral *m*, coralino
cordillera cordillera
core testigo, núcleo
core sample muestra de núcleo
Coriolis effect efecto de Coriolis
corner rincón *m*, esquina

11

cornice cornisa

corrasion corrasión

corridor pasillo

corrode desgastar

corrosive corrosivo

corundum corindón

coulee coulee *m*, colada, zanja de desagüe, cañada

count contar

country rock roca madre

course rumbo

cove ensenada

covered cubierto

crack agrietar, grieta, rajadura

crag risco, peña, peñasco

crater cráter *m*

craton cratón *m*

creek arroyo, arroyuelo

creep desplazamiento, reptación

crest cresta

Cretaceous Cretácico

crevasse grieta

crevice grieta

crinoids crinoideos

crop out aflorar

cross-bedding estratificación cruzada, estratificación transversal

cross-section sección transversal

crucible crisol *m*

crush aplastamiento, aplastar, triturar

crust corteza

crustal cortical, cortezal

cryology criología

crystal cristal *m or adj*

crystalline cristalino

crystallography cristalografía

cube cubo

cuprite cuprita

current corriente *f*

curve curva

cusp orla, cúspide *f*

cut cortar, corte *m*, desmonte *m*

cutbank banco erosionado

cutoff resección

cycle ciclo

cyclone ciclón

cylinder cilindro

dacite dacita

dale vallecito

dam presa, represa

damp húmedo

dampen amortiguar

dark oscuro *adj*

Darwinian Darwiniano

Darwinism Darwinismo

data datos *np*

date fechar, datar, fetcha *n*

datum dato, datum *m*

debris escombros, rocalla, detrito

debris avalanche avalancha de detritos

debris slide deslizamiento de detritos

decay decaer

declination declinación

declivity declive *adj*

deep profundo *adj*, hondo *m* or *adj*

deepen profundizar

deep-sea profundidad oceánica, de altura

deformation deformación

degree grado

dell cañada

delta delta

deluge diluvio

dendritic dendrítico

dendrochronology dendrocronología

dense espeso

density densidad

deposit depósito, yacimiento

deposition deposición

depositional deposicional

depression depresión

depth profundidad

descent bajada

desert desierto

desert pavement pavimiento del desierto

desert varnish barniz del desierto

desiccate desecar

detachment fault falla de desgarre, falla de desprendimiento

detritus detrito

deviation desviación
Devonian Devónico
dextral fault falla dextral, falla a la derecha
diabase diabasa
diagenesis diagénesis *f*
diagonal diagonal *f*
diameter diámetro
diamond diamante *m*
diapir diapiro *m*
diapiric diapírico
diastrophic diastrófico
diastrophism diastrofismo
diatomaceous diatomáceo
diatoms diatomea
differential diferencial
differentiated diferenciado
difficult difícil
digital digital
dike dique *m*
dilation dilatación
dilute diluir
diluvium diluvión *m*
dinosaur dinosaurio
diopside diópsido
diorite diorita

dip inclinación, buzamiento, inclinar, buzar
dip angle ángulo de inclinación
dip fault falla de inclinación
direction rumbo
dirt tierra
discontinuity discontinuidad
discordance discordancia
dislocation dislocación
displacement desplazamiento
dissected peneplain peneplanicie encañada
dissection disección
disturbance perturbación
ditch zanja, fosa
diversion desviación
divide divisoria
dolomite (mineral) dolomita
dolomite (rock) dolomía
dome domo
dormant durmiente, latente
dorsal dorsal *adj* or *f*
down abajo
downcut excavar hacia abajo

downdip inclinación abajo, buzamiento abajo
downhill cuesta abajo
downstream río abajo
downthrown movido hacia abajo
downthrown side lado caido
downwarp alabeo hacia abajo
downwarp basin cuenca de alabeo
drainage drenaje *m*, desagüe *m*
drainage basin cuenca de drenaje
draw diseñar
drawing dibujo *n*
drift deriva, derivar
drifting sand arena movediza
drill barrenar, taladrar, perforar, taladro *n*
drill core nucleo de pozo, testigo
driller perfador
drilling mud lodo de perforación
drought seqía

drown inundar, ahogar
drowned valley valle inundado
drumlin drumlin *m*
dry diggings placer eluvial *m*
dry lake lago seco, playa
dump basurero, escombrera
dune duna, médano
dunite dunita
dust polvo
dynamic dinámico
dynamics dinámica

— E —

early temprano
earth tierra
earthquake temblor *m*, sísmo, terremoto
earthy terroso
east este
ebb reflujo
ebb and flow flujo y reflujo
echelon escalón *m*
echeloned fault falla escolonada
eclogite eclogita

eddy remolino

edge margen *m*, borde *m*

edge, on edge de canto

efflorescent eflorescente

effusion efusión

eflorescence eflorescencia

eject expeler, eyectar

ejecta eyecto

ejectum eyecto

Ekman spiral espiral de Ekman *f*

elastic elástico

elastic rebound rebote elástico *m*

elevate levantar

elevation elevación, levantamiento

ellipsoid elipsoide *m*

elongate alargar

elongation alargamiento

elutriation elutriación

embankment terraplén *m*

embayment bahía, ensenada

embed empotrar

emerald esmeralda

emerge emerger, salir, surgir

emergent emergente

emery esmeril *m*

endogenous endógeno

en echelon en escalón

en echelon fault falla escolonada

energy energía

engineer ingeniero

enormous enorme

enrichment enriquecimiento

entrenched meander meandro intrincherado

entrance acceso

environment medio ambiente, cercanías

Eocene Eoceno

eolian eólico

eon eón *m*

epeirogenic epeirogénico

epeirogeny epeirogenía

ephemeral efímero

epicenter epicentro

epigene epigénico

epigenesis epigénesis *m*

epigenetic epigénetico

episode episodio

episodic episódico
epoch época
Epsom salt sal de Epsom *f,* epsomita
epsomite epsomita, sal de Epsom
equator ecuador *m*
equatorial ecuatorial
era era
erode desmoronar, erosionar
erosion erosión
erratic errático *m* or *adj*
eruption erupción
eruptive rock roca eruptiva
escarpment escarpa, barranca
esker esker *m*
estuary estuario
evaporate evaporar
evaporation evaporación
evaporite evaporita
evapotranspiration evapotranspiración
event acontecimiento
evolution evolución
excavate excavar
exfoliate exfoliar

exfoliation exfoliación
exhumation exhumación
expansion expansión
experimental experimental
exploration geology exploración geológica
explosion explosión
exposure exposición
exsolution exsolución
extention extensión
exterior exterior *m* or *adj*
extinct volcano volcán apagado, volcán extinto
extrude expeler
extrusion extrusión
extrusive extrusivo
extrusive rock roca extrusiva

— F —

face frente *f,* cara
facet faceta, cara
facies facies *fs,* faz *f*
failure fracaso
fall derrumbe *m*
fall line linéa de cataratas por una meseta

17

fanglomerate fanglomerado
fault falla *m*
fault block bloque de falla *m*
faulted fallado
faulting fallamiento
fault line línea de falla
fault plane plano de falla
fault trace traza de falla
fault zone zona de falla
fauna fauna
feldspar feldespato
feldspathic feldespático
felsic félsico
ferric férrico
ferrite ferrita
ferromagnesian ferromagnesiano
ferrous ferroso
field campo
fill relleno, terraplén *m*
filter filtro
fire fuego
fissile físil, hendible, rajadizo
fissure fisura, grieta, barranco
flagstone losa, laja
flake escama, copo

flaky escamoso
flank flanco
flat llano, plano
flatness aplanamiento
flatten achatar
flint pedernal *m*, piedra de chispa
floe témpano, bandejón
flood inundar, inundación, diluvio, riada, avenida
flood tide pleamar
floor piso, fondo
flora flora
flow flujo *m* or *adj*
flow volume (river) caudal *m*
fluid fluido
flume caz *f*
fluorescence fluorescencia
fluorescent fluorescente
fluoride fluoruro
fluorine flúor *m*
fluorite fluorita
fluvial fluvial
fluvioglacial fluvioglacial
focus foco
fog niebla

fold pliegue *m*, plegar, doblar
folded plegado
folding plegamiento
foliate foliar
foliation foliación
fools' gold oro de tontos
foothill pie de montaña *m*, estribación *f*
footpath senda, sendero
footprint huella
footwall muro de base *f*
foraminifera foraminíferos
force fuerza
ford vadear, vado
foredeep antefosa
foreshock temblor previo, aviso sísmico
foreshore playa de la marea
forest selva, bosque
forked lightning rayo ahorquillado
formation formación
fossil fósil *m* or *adj*
fossiliferous fosilífero

fossilization fosilización
fossilize fosilizar
founder hundirse
fountain fuente *f*
fractional fraccionario
fracture fractura
fragment fragmento, trozo
freeze congelar, helar
friable desmenuzable, friable
friction fricción
fringe orilla, borde *m*, margin *m*
front frente *f*, delantero *adj*
frontal delante, de frente
frost escarcha, helada
frost heave dilatación por congelación, fractura por congelamiento
fulgurite fulgurita
fuller's earth tierra de fuller
fumarole fumarola
furrowed surcado

— G —

gabbro gabro

19

galena galena
gallery galería
gangue ganga
gap vacio, desfiladero
garnet granate *m*
gas gas *m*
gem gema
genetic genético
geochemistry geoquímica
geochronology geocronología
geode géoda, drusa, bocarrena
geodesic geodésico
geodesy geodesía
geodetic geodético
geographic geográfico
geography geografía
geoid geoide *m*
geologic column columna geológica
geologic map mapa geologico *m*
geologic profile perfil geológico
geologic record registro geológico
geologic report informe geológico
geologic time tiempo geológico
Geological Survey Servicio Geológico
geologically geológicamente
geologist geólogo
geology geología
geomagnetic geomagnético
geomorphology geomorfología
geophysics geofísica
geosyncline geosinclinal *m*
geothermal geotérmico
geyser géiser *m*
glacial glacial, glaciario
glacial advance avance glacial
glacial drift deriva glacial, terrenos de acarreo
glacial outwash detrito glacial
glacial recession retroseso glacial

glaciation glaciación, helamiento, congelación
glacier glaciar *m,* helero
glass vidrio
glassy vítreo
global mundial
Global Positioning System GPS sistema de posicionamieto global *m*
globe globo
globigerina globigerina
globular globular
gneiss neis *m,* gneis *m*
gold oro
Gondwana Gondwana
gorge cañada, garganta, pongo, barranco, desfiladero
gouge jaboncillo, guibia
graben graben, zanja geológica, fosa tectónica
grade inclinación *f,* cuesta
gradient gradiente *f,* pendiente *f*
grain grano
grain size tamaño de grano

granite granito
granitic granítico
granodiorite granodiorita
granular granular *adj*
granularity granularidad
granulate granular *v*
granule gránulo
graphite grafito
graptolites graptolitos *mp*
grave huesa
gravel grava, cascajo
gravitation gravitación
gravitational force fuerza gravitacional
gravity gravedad
gravity anamoly anomalía de gravedad
gray gris
green verde
grid cuadrícula, reticulado
grind moler, triturar
grit cascajo, arena, gravilla
gritty arenoso
groove surco, acanaladura
grooved surcado
grotto gruta

ground suelo, tierra
ground water agua subterranea, agua freática
ground water recharge recarga de agua subterránea
grus grus *m*
gulch barranco, arroyo, quebrada
gulf golfo
gully barranco, hondonada
guyot guyot *m*
gypsum yeso
gypsum deposit depósito de yeso, yesera
gyre giro

— H —

hail granizo
half mitad, medio
half-life vida media
halite halita
halo halo
halogen halogéno
hammer martillo
hand lens lupa
hand level nivel de mano

hang colgar
hanging valley valle colgante *m*
hanging wall muro colgante
hard duro
hard hat casco
hardness dureza
hardpan caliche *m*, capa dura, tosca
headland promontorio
headwaters cabecera de rio
heat calor
heave levantamiento
heavy pesado
height altura
hematite hematita, hematites *fs*
hervidero volcán de lado
hiatus hiato
highland tierras altas, montañas
hightide pleamar, marea alta
highway carretera
hill loma, colina, cerro, cuesta

hinge charnela, bisagra

historical geology geología histórica

hogback cresta afilada, cuchilla, espinaza

hole agujero, hoyo, foso, fosa *m* or *adj*

hollow hueco

Holocene Holoceno

homocline homoclinal *m*

homogeneous homogéneo

hoodoo vea monigote, pilar de roca *m*

horizon horizonte *m*, capa

horizontal horizontal

hornblende hornablenda

hornfels hornfelsa, cornubianita, corneana

horst horst *f*, meseta tectónica

hot caliente

hot spot punto caliente

hot spring manantial de agua termal *m*, manantial caliente *m*

hour hora

huge enorme

humic húmico

hummock montículo, montecillo

hump montecillo

humus humus *m*

hurricane huracán *m*

hybrid híbrido

hydrate hidrato *m*, hidratar

hydration hidratación

hydraulic hidráulico

hydrite hidrita

hydrocarbon hidrocarburo

hydrochloric acid ácido clorohídrico

hydrology hidrología

hydrostatic hidrostático

hydrothermal hidrotermal

hydrothermal deposit depósito hidrotermal

hydrothermal vent chimenea hidrotermal

hydrous acuoso

hypsometric hipsométrico

ice hielo

ice age época glacial

iceberg iceberg *m*, témpano de hielo

ice cap calota de hielo

ice floe témpano de hielo

Iceland spar espato de Islandia

icepack bloque de hielo *m*

ice sheet capa de hielo, cubierta de hielo, manto de hielo

igneous ígneo

igneous body cuerpo ígneo

igneous rock roca ignea

ignimbrite ignimbrita

illuvial iluvial

illuviation iluviación

illuvium iluvio

ilmenite ilmenita

impact crater cráter de impacto *m*

impactite impactita

impermeable impermeable

incised inciso

incised meander meandro cortado, meandro excavado

inclination inclinación, declive *m*, buzamiento

incline inclinar

inclusion inclusión

incompetent incompetente

index índice

index fossil fósil índice

indurate indurar

induration induración

initial inicial

inject inyectar

injection inyección

inlet ensenada

inlier relicto interior

inner core núcleo interior

insect insecto

inselberg montaña isla, inselberg *f*

inshore hacia la orilla *adv*, cerca de la orilla *adv*, cerca de la costa *adv*

insolation insolación

insoluble insoluble

intensity intensidad
interbedded interestratificado
intercalate intercalar
intercalation intercalación
interglacial interglacial
interior interior *m* or
    adj
interlobate interlobulado
intermediate intermedio
intermittent intermitente,
    transitorio
intermontane intermontano
internal interno
interstage interetapa
interstate interestatal
interstate highway
    autopista
interstice intersticio
intersticial intersticial
interval intervalo, espacio
intricate intricado
intrusion intrusión
intrusive intrusiva
intrusive rock roca intrusiva
inundate inundar
inundation inundación

invasion invasión
invasive invasor
inverse inverso
inversion inversión
invert invertir
inverted invertido
iridium iridio
iron hierro
iron ore mena de hierro
iron ore hematite hematita,
    mena de hierro
isinglass cola de pescado
island isla
island arc arco de islas
isochrone isócrona
isocline isoclinal *m*
isomorph isomorfo
isomorphic isomórfico
isopach isopaca
isostacy isostasia
isostatic isostático
isotope isótopo
isotopic isotópico
isotropic isotrópico
isotropy isotropía
isthmus istmo

— J —

jade jade *m*
jadeite jadeita
jagged mellado, dentado
jasper jaspe *m*
jet stream corriente en chorro *f*
joint diaclasa, juntura
jointed diaclasado
jointing diaclasamiento
jungle selva, jungla
Jurassic Jurásico

— K —

kame kame *m*
kaolin caolín *m*
kaolinite caolinita
kaolinization caolinización
karst cárstico *adj*, carso *n*
kettle hoyo glaciario
key (reef) cayo
kimberlite kimberlita
knife cuchillo, machete
knob protuberancia
knoll otero
knot nudo

kyanite cianita

— L —

labradorite (mineral) labradorita
labradorite (rock) labrador *m*
laccolith lacolito
lacustrine lacustre
lag retardo, rezagarse
lagoon laguna
lahar lahar *m*
lake lago
lamella laminella
lamina lámina
laminated laminado
land tierra
landform forma de relieve, forma fisográfica
landmass masa terrestre, tierra firme
landscape paisaje
landslide deslizamiento de tierra, corrimiento de tierra, derrumbamiento de tierra
lapis lazuli lapis lázuli *m*

large grande
large rock peñasco, roca grande
late tardío
lateral lateral, lado
lateral moraine morrena lateral
laterite laterita
lateritic laterítico
latitude latitud *f*
lattice red
Laurasia Laurasia
lava lava
lava flow colada
layer estrato, capa, lecho, cama
layered estratificado
leach lixiar
lead plomo
lean delgado
ledge retallo
lee sotavento *m o adj*
left-lateral fault falla lateral izquierda
length longitud *f*
lens lente *f*

levee levee *f,* dique *m*
level nivel *m*
lift levantar
light luz *n,* ligero *adj*
lightning relámpago
lightning flash rayo
lignite lignito
lime cal *f,* lima, calizo
limestone caliza *f*
limonite limonita
lineal lineal
lineament lineamento
linear linear, lineal
liquid líquido *m* or *adj*
listric lístrico
lithic lítico
lithification litificación
lithified litificado, petrificado
lithium litio
lithografic litográfico
lithography litografía
lithology litología
lithosphere litosfera
lithostatic litostático
littoral litoral *m* or *adj*
littoral drift deriva litoral

load peso
loam marga
lobate lobulado
lobate moraine morrena lobulada
lobe lóbulo
lode filón, veta
lodestone piedra imán, magnetita
loess loess *m*
long largo
longitude longitud *f*
long runout (landslide) de gran alcance, delizamiento, sturzstrom
longshore costero
loose suelto
lower más bajo
low tide bajamar
low velocity zone zona de baja velocidad
loxodrome loxodromía

— M —

maar maar *m*
mafic máfico, ferromagnesiano

magma magma *m*
magmatic magmático
magmatic chamber cámara magmática
magnesia magnesia
magnesian magnesiano
magnesite magnesita
magnesium magnesio
magnet imán *m*
magnetic field campo magnético
magnetism magnetismo
magnetite magnetita
magnify aumentar, ampliar
magnifying glass lente *f* de aumento, lupa
malachite malaquita
mamelon mamelón
mammoth mamut *m*
manganese manganeso
manganese nodule nódulo de manganeso
mantle manto
map mapa *m*, carta
marble mármol *m*
margin margen *m*

marginal marginal
marine marino
mark marca, marcar
marl marga, greda
marsh ciénega, pantano
massif macizo *m*
mass masa
mass wasting remoción en masa, degaste *m*
massive macizo
matrix matrix
matter materia
mature maduro
maturity madurez *f*
meadow prado
mean promedio
meander meandro
mechanical mecánico
mechanics mecánica
medial medial
medial moraine morrena intermedia
median mediana
mediterranean mediterráneo
medium mediano
melt fundir, derretir

meltwater aguahielo
member miembro
mercury mercurio
meridian meridiano
meridional meridional
mesa mesa
mesh red
mesosphere mesósfera
Mesozoic Mesozóico
metal metal *m*
metallic core nucleo metálico
metamorphic metamórfico
metamorphic rock roca metamórfica
metamorphism metamorfismo
meteor meteoro
meteorite meteorito
meteoritic meteórico
meteorology meteorología
methane metano
mica mica
micaceous micáceo
micaschist micacita
microbreccia microbrecha

microcline microclina
microcrystaline microcristalino
microfossil microfósil *m*
microscope microscopio
middle medio
mid-ocean ridge dorsal centro oceánica *f*
mill moler
mine mina
mineral mineral *m* or *adj*
mineralization mineralización
mineralize mineralizar
mineral resources recursos minerales
minerology mineralogía
mining minería
minute minuto
miocene mioceno
Mississippian Misisípico
Missourian Misuriense
mist neblina
mixed mixto
Mohs scale escala de Mohs
moisture humedad
molecule molécula

mollusc molusco
molten fundido
molybdenum molibdeno
monadnock monadnock *m*
monazite monacita
monocline monoclinal *m*
monsoon monzón *m*
montmorillonite montmorillonita
monzonite monzonita
moon luna
moonstone albita
moraine morrena
mound montículo
mount monte *m*
mountain montaña
mountainous montañoso
mountain range cordillera
mouth boca
mud barro, lodo, limo, lama, cieno, fango
muddy lodoso
mudflat marisma
mudflow corriente de lodo *f*, corriente del barro *f*, flujo de lodo

mudstone roca arcillosa, lutolita
mud volcano volcán de lodo *m*
muscovite muscovita
mylonite milonita
mylonitic milonítico
mylonitization milonitización

— N —

narrow estrecho
native nativo
natural natural
natural arch arco natural
natural bridge puente natural *m*
natural gas gas natural
natural selection selección natural
nature naturaleza
nautical náutico
nautiloids nautiloideos
neap tide marea muerta
neck cuello
needle aguja
negative negativo

nekton necton *m*
Neogene Neogeno
nepheline nefelina
nephelinite nefelinita
nephrite nefrita
net red
nickel níquel *m*
nitrate nitrato
nitrogen nitrógeno
nivation nivación
node nodo
nodule nódulo
normal normal
normal fault falla normal
north norte *m* or *adj*
North American Datum 1927 NAD27 Datum Norteamericano de 1927
northern septentrional
nose nariz *f*
notebook cuaderno *f*
now ahora
nuee ardiente nube ardiente *f*
nugget pepita
nunatak nunatak *m*, pico solitario *m*

31

oasis oasis *m*
oblate oblado
oblique oblicuo
obsidian obsidiana
obtuse obtuso
ocean océano
oceanic oceánico
ocher ocre *m*
offset separación
offshore costa afuera, mar
    adentro
oil petrólio, aceite
oil sand arena petrolifera
oil shale petrolutita
old senil, viejo
Oligocene Oligoceno
oligoclase oligoclasa
olivine olivino
onyx ónix *m*
oolite oolita
oolith oolita
ooze filtrar, cieno, lama,
    fango
opal ópalo
opaque opaco

open abierto
open cut corte abierto,
    rajadura a cielo abierto,
    tajo a cielo abierto, rasgo
    abierto
open pit mine mina a cielo
    abierto
open trench trinchera
    abierta, rasgo abierto
ophiolite ofiolita
orange anaranjado
orbit órbita
Ordovician Ordovícico
ore mena
orient orientar
origin origen
orogene orógenio
orogenic orogénico
orogeny orogenia,
    orogénesis *m*
orographic orográfico
orthoclase ortoclasa
oscillate oscilar
osmium osmio
outcrop afloramiento, asomo
outflow efusión, desagüe *m*

outlet salida, escape *m*, desembocadura, desagüe *m*

outlier relicto exterior, roca apartada

outpouring derrame *m*

outwash deslava, alluvión *m*

oven horno

overburden sobrecapa, cubierta

overflow derrame *m*

overhang sobrecolgante, sobresalir, cornisa

overlap solapar, traslapar, traslapo *m*

overlapping solapamiento

overlie suprayacer

overlying suprayaciente

overmass masa suprayaciente

overthrust sobrecorrimiento, sobrecorrido

overthrusted sobreempujado

overturned invertido

oxbow lake laguna semilunar, meandro abandonado, brazo muerto

oxidation oxidación

oxide óxido

oxidize oxidar

oxygen oxígeno

ozone ozono

— P —

P wave onda P

pace paso

packet paquete

packing density densidad *f* de empaque *m*

pahoehoe pahoehoe, lava cordada, lava fluida

paladium paladio

Paleocene Paleoceno

Paleogene Paleogéno

paleontology paleontología

paleosol pleosol m

Paleozoic Paleozoico

palinspastic palinspastico

pampa pampa

Pangea Pangea

paralic parálico

parallel paralelo

particle partícula

pass paso
passage pasaje
path vereda, senda, sendero
pavement pavimento, calzada
peak pico, cumbre *f*, cima
peat turba
pebble guija
pediment pedimento
pedology pedología
pegmatite pegmatita
pelagic pelágico
Pelean Peleano
pellet pelotilla
pen pluma
pencil lápiz
peneplain peneplanicie
peninsula peninsula
Pennsylvanian Pennsilvánico
perch percha
perched rock roca
   encaramada
percolate percolar, infiltrar
percolation percolación,
   filtración
perfect perfecto
perforate perforar

peridot peridoto
peridotite peridotita
period período
perlite perlita
permafrost pergelisol *m*
permeable permeable
Permian Pérmico
perspective perspectiva
pervious permeable
petrified petrificado
petrify petrificar
petroleum petróleo
petrology petrología
phaneritic rock roca fanerítica
phase fase
phenoclast fenoclasto
phenoclastic fenoclástico
phenocryst fenocristal *m*
phosphate fosfato
phosphorite fosforita
photograph fotografía
phreatic freático
phreatophyte freatofito
physical físico
physics física
physiographic fisiográfico

piece pedazo, pieza
piedmont piamonte *m*
pillar pilar *m*
pillow lava almohadilla de lava
pink rosado
pink agate cornalina
pinnacle pico, cumbre
pipe tubo, tubería, chimenea
pit hoyo, foso
pitch inclinación, brea
pitchblende pechblenda, uraninita
pitted hoyoso
place, in place en sitio
placer placer *m*
placer deposit deposito de placer *m*
plagioclase plagioclasa
plain planicie *f*, llanura, llano *m*
planation rebajamiento, aplanamiento
plane plano *m* or *adj*
planet planeta *m*
plane table plancheta

planetesmal planetesmal *m*
plankton plancton *m*
plant planta
plaster yeso
plasticity plasticidad
plate placa
plateau meseta, altiplano
plate tectonics tectónica de placas
platform plataforma
platinum platino
platy laminado
playa playa, lago seco
playa lake laguna efímera, playa efimera
Pleistocene Pleistoceno
Pliocene Plioceno
pluck extraer
plug tapón, obturador *m*
plugged obturado
plunge buzar
plunging buzante
pluton plutón *m*
plutonic plutónico
pluvial pluvial
pocket bolsillo

point bar barra de punta
point (of) punta
point (to) punto
polar polar
polar wandering deriva
   polar
pole polo
pollen polen *m*
polyconic policónico
polyp pólipo
pond estanque *m*
pool poza
porosity porosidad
porous poroso
porphyry pórfido
porphyritic porfirítico
positive positivo
post poste
postglacial posglacial
potash potasa
potassium potasio
powder polvo
prairie pradera
Precambrian Precámbrico
Precambrian platform
   plataforma Precámbrica

precipice precipicio,
   despeñadero
preservation preservación
principle principio
primary primario
presently ahora
pressure presión
profile perfil *m*
projection proyección
promontory promontorio
prospect prospectar
Proterozoic Proterozoico
province provincia
proximity proximidad
pseudomorph pseudomorfo
pull tirar
pulverize pulverisar, triturar
push empujar
pumice pómez *m* or *f*,
   pumicita
pyrite pirita
pyroclastic piroclástico
pyroxene piróxena

—Q—

quarry cantera

36

quartz cuarzo
quartz bearing cuarcífero
quartzite cuarcita
quartzose cuarzoso
Quaternary Cuaternario
quicklime cal viva
quicksand arena movediza

— R —

radioactive decay
decaimiento radioactivo
radioactivity radioactividad
radiolaria radiolarios
radiometric dating
fechamiento radiométrico
(isotópico)
radius radio
radon radón *m*
raft balsa
railway ferrocarril *m*
rain lluvia, llover
rainbow arco iris
raise levantar
range sierra, cordillera,
alcance *m*
rank rango

rapidity rapidez
rapids correntera, rápidos
rare earths tierras raras
rate razón, valuar *m*
ratio relación, proporción
ravine barranco, cañada,
garganta, quebrada
ray rayo
razorback cuchilla
reach alcance *m*
rebound rebote *m*
recent reciente
recession retroceso, retirada
recessional moraine morena
de retroceso
reconnaissance
reconocimiento
reconnoiter reconocer
record registro
recovered recuperado
recovery recuperación
rectangle rectángulo
recumbent fold pliegue
recostado *m*, pliegue
reclinado
red rojo

reef arrecife *m,* escollo, bajio, cayo

reference level nivel de referencia

reflection reflexión

refraction refracción

regional dip buzamiento regional

regional metamorphism metamorfismo regional

regolith regolito

regression regresión

rejuvenation reactivación, rejuvenecimiento

release fault falla de relajación

relic sea mar antiguo

relict relicto

relief relieve *m*

remove retirar

replacement reemplazo

report informe

reservoir yacimeinto, cisterna

residual residual

resin resina

resistant resistente

resistence resistencia

retreat retroceso

retrocessional moraine morena de retroceso

reversal reversión

reverse fault falla inversa

review revisar, repasar

rhodochrosite rodocrosita

rhombic rómbico

rhyolite riolita

rice paddy arrozal

ridge cresta, dorsal *f*

rift hendidura

rift valley valle tectónico

right angle ángulo recto

right-lateral fault falla lateral derecha

rim borde *m*

rip current corriente de retorno *f*

ripple rizadura

rise levantamiento

river río

river bank ribera

riverbed lecho, cauce *m*

river mouth
desembocadura de rio
road camino
roche moutonee roca
aborregada
rock roca, peña
rockfall derrumbamiento de
rocas
rocky rocoso, pedregoso,
pétreoso, cantoso,
peñascoso
roof pendant colgajo
rose quartz cuarzo rosado
rounded redondo
rounding redondeamiento
roundness redondez *f*
route ruta, recorrido
rubble escombros, ripio
rubidium rubidio
ruby rubí *m*
rupture ruptura
rust orín *m*, herrumbre *f*,
enmohercerse
rusty herrumbroso
rutile rutilo

— S —

S wave onda S
saddle silla
safe seguro
sag pond laguna de
desplome
salina salina
saline salino *adj*
salt sal *f*
salt crystal sal gema
salt dome domo salino
salt flat salina
salt mine salina
saltpeter salitre *m*, caliche *m*
salt water auga salada
salty salado
sample muestra, muestrear,
recolectar
sand arena
sandbank banco de arena
sandbar barra de arena,
tómbolo
sandblast lijar con arena
sandstone arenisca
sandy arenoso
sandy clay arcilla arenosa

sapphire zafiro
sapping socavación
satellite satélite *m*
savanna sabana
scabland terreno costroso
scale escala
scaly escamoso
scarp escarpa
schist esquisto
scissors fault falla pivotal
scoria escoria
scour (channel) limpiar
scratch rasguño
scree escombro de talud,
    pedrero, derrubio
screen acribar
sea mar *m* or *f*
sea cliff acantilado
seafloor spreading expansión
    de los fondos oceánicos
sealed obturado
sea level nivel de mar
seam veta
seamount monte submarino
    m, pico submarino
season estación

sea stack farallón *m*
sea urchins erizos del mar
secondary wave onda
    secundaria
section sección
secure seguro
sediment sedimento
sedimentary sedimentario
sedimentary rock roca
    sedimentaria
sedimentation sedimentación
sedimentology
    sedimentología
seep filtrarse, rezumarse
seepage filtración, sobrencia
seismic sismico
seismic reflection reflexión
    sísmica
seismic refraction
    refracción sísmica
seismic seawave onda
    sísmica marina
seismic wave onda sísmica
seismograph sismógrafo
seismologist sismólogo
seismology sismología

seismometer sismómetro
selenite selenita
selenium selenio
semiarid semiárida
senility senectud *f*
sequence secuencia, sucesión
serial serie *f*
series serie *f*
serpent serpiente
serpentine serpentina
serration serración
sessile sésil, sentado
settlement asentamiento
shake sacudir, sacudida *f*
shale lutita
shallow somero, poco profundo
shallow sea mar playa
shallow water aguas someras
shaly lutítico
shard pedazo, triza
sharp afilado, agudo
sharpening stone piedra de afilar
shatter triturar, hacer añicos
shear cizallar, cizalla

shear fault falla de cizalla
shearing cizallante
sheet manto, sábana, hoja de papel, extensión, lámina
shelf plataforma
shell concha, valva
shield escudo
shield volcano volcán en escudo *m*
shingle grava de playa, teja
shoal bajío, banco de arena
shock choque *m*, conmoción
shock wave onda de choque
shore orilla
sial sial
side lado *n*, lateral *adj*
sidereal sidéreo
siderite siderita
sierra sierra
sieve acribar, tamiz *m*
sift acribar
silex cuarzo pulverisado, chert
silica sílice

41

silicate silicato

siliceous silíceo

silicon silicio

silicification silificación

sill silo

silt limo, cieno

silty limoso

Silurian Silúrico

silver plata

sima sima

similar similar

sine sino

sink depresión carsítica, hundirse

sink hole agujero de desagüe, sumidero

sinter toba

situ, in situ en sitio

skarn skarn *m*

skeleton esqueleto

sketch dibujo

skinny flaco

sky cielo

slab losa, laja, plancha, tabla

slag escoria

slate pizarra

slaty pizarroso

slice rebanada

slickensides superficie de deslizamiento, espejo de fricción

slide desliz *m*, deslizamiento

slide rule regla de cálculo

slime cieno, lama, limo

slip desliz *m*, deslizamiento, resbalamiento

slip-off slope pendiente de deslizamiento *f*

slippery resbaloso

slope talud *m*, declive *m*, cuesta, inclinación, pendiente *f*

slope wash lavaje *m* de pendiente

slough cenegal *m*, légamo

sludge ice capa de hielo nuevo, hielo pastoso

slump derrumbe *m*

slurry lechada, compuesto acuoso, pasta aguada

smooth liso *adj*, alisar, suave *adj*

snail caracol *m*

snake culebra, serpiente *f*

snow nieve *f*, nevar

snowstorm nevasca

soda soda

sodium sodio

soft suave, blando

soil suelo, tierra, terreno

soil mechanics mecánica de suelos

soil sample muestra de suelo

solid sólido *m* or *adj*

soluble soluble

solute soluto

solution solución

sort seleccionar, clasificar

sorted clasificado

sorting selección

sounding sondeo

source fuente *f*

south sur

southern meridional

spar espato

sparkling brillante

species especie *f*

specimen muestra, ejemplar

speed rapidez, velocidad

speleology espeleología

sphere esfera

spherical esférico

spheroid esferoide *m*

spillway vertedero

spinel espinela

spiral espiral *m* or *adj*, carocol *m* or *adj*

spit lengua de tierra, unida a la costa

spodumene espodumeno

spout tromba

spread esparcir

spring manantial *m*, fuente, vertiente, ojo de agua

spur estribación, estribo

square cuadrado

stable estable *adj*

stage etapa

stain mancha

stalagmite estalagmita

stalagtite estalactita

standards normas

star estrella
steel acero
steep escarpado, empinado
stereo pair par estereoscópico
sticky pegajoso
stone piedra
stony pétreo
stope excavación para minar
straight derecho
strain deformación
straits estrecho, pasaje
strata estratos, lechos, capas
stratification estratificación
stratified estratificado
stratigraphic estratigráfico
stratigraphy estratigrafía
stratosphere estratosfera
stratovolcano estratovolcán *m*
stratum estrato, capa, cama
streak raspadura
stream arroyo
stream meandering cauce meándrico, corriente meándrica
strength resistencia
stress estrés *m*, estresan

stria estría *fs*
striae estrías *fp*
striate estriar
striation estriación
strike rumbo del estrato
strike of fault rumbo de la falla
strike-slip fault falla de rumbo, falla de desplazamiento de rumbo
strip faja, arrasar
strip mine mina a cielo abierto
stromatolite estromatolito
strong fuerte
strontium estroncio
structural control control estructural *m*
structural feature rasgo estructural
structural geology geología estructural
structure estructura
sturzstrom sturzstrom
subduct subducir
subduction subducción

44

submarine submarino
submarine earthquake
maremoto
submarine fan abanico
submarino
submarine trench fosa
oceánica, trinchera
submarina, fosa
submarina
subsea bajo el nivel del mar
subsequent subsecuente
subside hundirse
subsidence hundimiento
subsoil subsuelo
subsurface subsuelo
sugarloaf pan de azúcar
sulfate sulfato
sulphur azufre *m*
sulphurous sulfuroso
summary sumario
summit cima, cumbre *f*
sun sol *m*
supercrustal supracortical
superimpose sobreimponer
surf rompiente *m*
surface superficie *f*

surveying topografía,
agrimensura
surveyor topógrafo,
agrimensor
suspend suspender
suspension suspensión
swamp ciénega, pantano
swampy cenagoso, pantanoso
swing ocilación
syenite sienita
symbol símbolo
symmetric simétrico
symmetry simetría
synclinal fold pliegue
sinclinal *m*
syncline sinclinal *m*
system sistema *m*

— T —

tailings relaves *mp*
talc talco
talus talud *m*
talus slope pendiente de
escombros *f*

tangent tangente
tar brea, alquitrán *m*
taxonomy taxonomía
tectonic tectónico
tektite tectita
telescope telescopio
temblor temblor *m*
temperature temeratura
tension tensión, tensionar
tephra tefra
terminal moraine morrena terminal
terrace terraza, terraplén
terrain terreno, topografía
terrane terreno, topografía
terrestrial terrestre
terrigenous terrígeno
Tertiary Terciario
Tethys Tetis
texture textura
theodolite teodolito
thermocline termoclina
thermosphere termosfera
thick espeso
thickness espesor *m*
thin delgado, flaco

thin section lamina delgada, sección delgada
thixotropic tixotrópico
tholeitic toleítico
thorium torio
throw desplazamiento vertical
thrust fault falla de corrimiento, falla inversa, falla de cabalgadura
thunder trueno
tidal bore ola de marea
tidal channel canal de marea *m*
tidal flat planicie de marea
tide marea
tight compacto *adj*
till till *m*
tillite tillita
tilt inclinarse, ladearse
time tiempo
time scale escala de tiempo
tin estaño
titanium titanio
tombolo tómbolo

46

tomography tomografía
tonalite tonalita
top cumbre
topaz topacio
topographer topógrafo
topographic topográfico
topography topografía
tornado tornado
torrent torrente *m*
tourmaline turmalina
trace traza, rastro
tracing caleo
trachyte traquita
track huella
trade winds vientos alisios
trade wind zones fajas de
  alisios
trail trocha, senda, pista,
  vereda
transform fault falla
  transformante
transgression transgresión
transit teodolito
transport transporte *m*
trap (rock) trap *m*
traverse atravesar, recorrer

travertine travertino
tremor temblor *m*
trench fosa, zanja, trinchera
trend rumbo, dirección
triangle triángulo
Triassic Triásico
trilobites trilobites *mp*
tripod trípode
triturate triturar
tropic trópico
tropical tropical
troposphera troposfera
trough depresión, artesa,
  seno, hoya
truck camión *m*
truncate truncar
tsunami tsunami *m*
tufa tufa
tuff toba
tuffaceous tobáceo
tumulus túmulo
tundra tundra
tunnel túnel *m*
turbidite turbidita
turquoise turquesa
twin gemelo

typhoon tifón

— U —

ultrabasic ultrabásico
ultramafic ultramáfico
unattached no fijo
unconformity discordancia
under debajo de
undercut socavar, socavado
undercut slope pendiente
   socavada
undercutting socavación
underlying subyacente
undermass masa infrayacente
undersea volcano volcán
   submarino
undertow resaca
undisturbed no perturbado
uniformitarianism
   uniformitarismo
Universal Tranverse
   Mercator UTM
   Universal Traversa de
   Mercator
unstable inestable
updip buzamiento arriba

uphill cuesta arriba
uplift levantamiento
upper más alto
upstream río arriba
upthrown levantado,
   sobrelevantado
upthrown side lado levantado
upthrust empuje *m*
   ascendente
upwelling surgencia,
   surgente
uranite uranita
uranium uranio
U-shaped valley valle en
   forma de «U»
UTM projection
   proyección UTM

— V —

valley valle *m*
vapor vapor *m*
variation variación
varve varva
vein veta, filón, vena
veinlet venilla
veinule venilla

velocity velocidad
vent chimenea, respiradero
ventifact ventifacto
vermicullite vermiculita
vertex vértice
vertical vertical
vesicle vesícula
vesicular vesicular
vesicular basalt basalto
vesicular
Vesuvius Vesubio
viscosity viscosidad
vitreous vítreo
vitriol vitriolo
volcanic volcánico
volcanic bomb bomba
volcánica
volcanic dome domo
volcánico
volcanic eruption erupción
volcánica
volcanic mud lodo volcánico
volcanic plug tapón volcánico
volcanic vent chimenea
volcánica
volcano volcán *m*

vortex remolino
V-shaped valley valle en
forma de «V»
vug vesicula, géoda

— W —

wacke wacka
wall (rock) muro, muralla,
pared
wandering of the poles
deriva migración de los
polos
warped alabeado
wash lavar
water agua
watercourse curso del agua
waterfall cascada, salto de
agua, catarata
water gap garganta
watershed divisoria,
vertiente *m*
waterspout tromba marina,
manga
water table nivel freático
watery aguoso
wave ola, onda

wave-cut cortada por las olas, erosionado de ola
wavelength longitud de onda
way sendero, camino
weak débil
weather desgastar, meteorizar, tiempo *n*
weathered meteorizado
weathering meteorización
wedge cuña
wedge-shaped cuneiforme
weight peso
welded pumice pumicita soldada, pómez soldada, ignimbrita
welded tuff toba soldada, ignimbrita
well pozo
west oeste
whirlpool remolino
whirlwind remolino
white blanco
wide ancho, amplio
wind viento
wind gap abra de viento
window ventana

windward barlovento *m*, de bartolovento *adj*
wolfram wolram *m*, volframio
woods bosque
wrinkle arrugar, arruga
wrinkled rugoso

— X —

xenolith xenolito

— Y —

yellow amarillo
yield ceder
young joven

— Z —

zeolite zeolita
zinc cinc *m*
zircon circón *m*
zone zona
zoolith zoolita
zoology zoología

Español - Inglés

aa aa

abajo down, below

abanico aluvial alluvial fan

abanico submarino submarine fan

abierto open

abisal abyssal

abismo abyss

ablación ablation

abra de viento wind gap

abrasivo abrasive

absorber absorb

acanaladura groove

acantilado sea cliff

acceso entrance

accidente *m* accident

aceite oil

acero steel

ácido acidic, acid

ácido clorohídrico hydrochloric acid

acondrito achondrite

acontecimiento event

acordanza conformity

acreción accretion

acribar sift, sieve, screen

acuático aquatic

ácueo aqueous

acuífero aquifer

acuoso aqueous, hydrous

achatar flatten

adobe *m* adobe

aéro aerial

aeróbio aerobic

aerosol *m* aerosol

afilado sharp

afloramiento outcrop

aflorar crop out

ágata agate

agregado aggregate

agrietar crack

agrimensor surveyor

agrimensura surveying

agua water

agua de fusión de la nieve meltwater

agua fósil connate water

agua freática ground water

aguahielo meltwater

agua salada salt water

aguas someras shallow water

agua subterranea ground water

agudo sharp

aguja needle

agujero hole

agusos watery

agujero de desagüe sink hole

ahogar drown

ahora now, presently

aire air

alabastro alabaster

alabeado warped

alabeo hacia abajo downwarp

alargamiento elongation

alargar elongate

albita albite, moonstone

álcali *m* alkali

alcalino alkaline

alcance range, reach

aleación alloy

alga alga

algáceo algal

algas algae

alidada alidade

alisar smooth

almohadillada de lava pillow lava

alóctono allochthonous

alpino alpine

alquitrán *m* tar

alquitrán de hulla *m* coal tar

alteración alteration

alterado altered

altímetro altimeter

altiplano plateau

altitud *f* altitude

altura height

altura, de altura deep sea

alud *f* avalanche

alud de detritos debris avalanche

alumbre *m* alum

alumina alumina

aluminio aluminum

aluvial alluvial

aluvión *m* alluvium

amalgama amalgam

amarillo yellow

amatista amethyst

ambár *m* amber

amígdala amygdule
amigdaloide *m* amygdaloid
amonite *mp* ammonite
amorfo amorphous
amortiguar dampen
ampliar magnify
amplio wide, broad
ampolla blister
anaerobio anaerobic
anamolía de gravedad
  gravity anamoly
anarnjado orange (color)
anastomosado braided
ancho wide
andesita andesite
anfíbol *m* amphibole
anfibolita amphibolite
angular angular
ángulo angle
ángulo de inclinación angle
  of dip, dip angle
ángulo de reposo angle of
  repose
ángulo recto right angle
ánguloso angular
anhidro anhydrous

anhidrita anhydrite
anisotropía anisotropy
anisotrópico anisotropic
anisótropo anisotropic
anomalía anomaly
anomalía de gravedad
  gravity anomaly
anómalo anomalous
anortita anorthite
anortoclasa anorthoclase
antárctico antarctic
Antártida Antarctica
antecedente antecedent
antefosa fore deep
anticlinal *m* anticline
antiguo ancient
antimonio antimony
antimonita antimonite
antracita anthracite
anual annual
apatita apatite
ápice apex
aplanamiento flatness,
  planation, crush
aplastar crush
aragonita aragonite

arcén berm
arcilla clay
arcilla arenosa sandy clay
arcillífero argilliferous
arcillita claystone
arcilloso argillaceous, clayey
arco arc, arch
arco de islas island arc
arco iris rainbow
arco natural natural arch
arcosa arkose
arcósico arkosic
arena sand, grit
arenáceo arenaceous, sandy
arena movediza drifting
   sand, quick sand
arena petrolifera oil sand
arenoso sandy, gritty
arenisca sandstone
argiláceo argillaceous
argillita argillite
argón *m* argon
árido arid
arquear arch
Arqueozoico Archeozoic
arrasar strip

arrecife *m* reef
arroyo arroyo, stream, creek,
   gulch
arroyuelo brook, creek
arrozal rice paddy
arruga wrinkle
arrugar wrinkle
arsénico arsenic
artesa trough
artesiano artesian
ártico arctic
asbesto abestos
asentamiento settlement
asfalto asphalt
asimetría asymmetry
asimétrico asymmetric
asomo outcrop
astenosfera asthenosphere
asteroide *m* asteroid
atmósfera atmosphere
atolón atoll
atravesar traverse
aumentar magnify
auroras aurora
austral austral
autóctono autochthonous

autopista interestatal interstate highway

avalancha avalanche

avalancha de detritos debris avalanche

avance glacial glacial advance

avenida flood

aviso sísmico foreshock

axial axial

azimut *m* azimuth

azufre *m* sulphur

azul blue

azurita azurite

— B —

bahía embayment, bay

bajada bajada, descent

bajamar low tide

bajío shoal, reef

bajo el nivel del mar subsea

balasto ballast

balso raft

banco bank, bench

banco de arena sandbank

banco en río bar

banco erosionado cutbank

banda band

bandeado banded

bandeando banding

bandejón floe

barita barite

barján barchan

barlovento *m* windward

barloviento, de barloviento windward

barniz del desierto desert varnish

barra bar

barra de arena sandbar

barra de punta point bar

barranca ravine, escarpment

barranco gully, ravine, gorge, fissure, gulch

barrenar drill

barrera barrier

barro mud, clay

basáltico basaltic

basalto basalt

basalto vesicular vesicular basalt

basamento basement

base base
básico basic
basurero dump
batimetría bathymetry
batolito batholith
batrimétrico bathymetric
bauxita bauxite
belemnites *mf* belemnites
bentonita bentonite
berilio berrylium
berilo beryl
berma berm
binario binary
biogénesis *f* biogenesis
biogénico biogenic
biología biology
bioquímico biochemical
biotita biotite
bisagra hinge
biselar bevel
bismuto bismuth
bivalvos bivalves
blanco white
blando soft
bloque *m* block, boulder
bloque de falla *m* fault block

bloque de hielo *m* icepack
boca mouth
bocarrena geode
bolsillo pocket
bolsón *m* bolson
bomba volcánica volcanic bomb
bórax *m* borax
borde *m* edge, rim
boreal boreal
boro boron
bosque forest, woods
botánica botany
botánico botanic
botanista botanist
botriodal botryoidal
botrioide *m* botryoid
braquiópodos brachiopods
brazo muerto oxbow lake
brea pitch, tar
brecha breach, breccia
brechoso brecciated
brillante bright, sparkling
brillo brightness
bronce *m* bronze
brotadero seep

brotante flowing
brújula compass
bufadora blowhole
burbuja bubble
buzamiento dip, inclination
buzamiento abajo downdip
buzamiento arriba updip
buzamiento regional regional dip
buzante plunging
buzar plunge, dip

— C —

cabecera de rio headwaters
cabo cape
cal *f* lime
cala bight
calcáreo calcareous
calcedonia chalcedony
calcificar calcify
calcio calcium
calcita calcite
calco tracing
calcopirita chalcopyrite
caldera caldera

caliche *m* caliche, hardpan, saltpeter
caliente hot
calita bight
caliza limestone
calizo calcareous, lime
calor heat
calota de hielo ice cap
cal viva quicklime
calzada pavement
cama layer, bed, stratum
cámara chamber, camera
cámara magmática magmatic chamber
Cámbrico Cambrian
camino road, way
camión *m* truck
campamento camp
campo field
campo magnético magnetic field
canal *m* channel
canal de marea tidal channel
cantera quarry
canto, de canto on edge
canto rodado boulder

cantoso rocky
cañada coulee, dell, ravine
cañón *m* canyon
caolín *m* kaolin
caolinita kaolinite
caolinización kaolinization
capa band, bed, horizon, layer, stratum
capa dura hardpan
capa de hielo ice sheet
capa de hielo nuevo sludge ice
caparazón carapace
capas strata
capaz capable
capilar *m* or *adj* capillary
capuchón cap rock
cara face, facet
caracol *m* snail, spiral
carapacho carapace
carbón coal
carbonato carbonate
carbón bituminoso bituminous coal
carbonífero carboniferous
carbono carbon

carbonoso carbonaceous
carburo carbide
carnotita carnotite
carretera highway
carso karst
cárstico karst
carta map, chart
cascada cascade, waterfall
cascajo gravel, grit
casco hard hat
casiterita cassiterite
cataclástico cataclastic
catarata cataract
catastrófico catastrophic
catastrofismo catastrophism
cauce *m* riverbed
cauce meándrico meandering stream
caudal *m* flow volume (river)
caverna cavern
cavidad cavity
cayo key, reef
caz *f* flume
ceder yield
cementación cementation

cementar *v* cement (to cement)

cemento *m* cement

cenagoso swampy

cenegal *m* slough

ceniza ash, cinder, sinter

Cenozoico Cenozoic

centígrado centigrade

centro center

ceolita zeolite

cerca de la costa inshore

cerca de la orilla inshore

cercano close

cercaniás environment

cerrado closed

cerrar close

cerro hill

cianita kyanite

ciclo cycle

ciclón cyclone

cielo sky

ciénega bog, marsh, swamp

cieno mud, ooze, slime, silt

cieno calcáreo calcareous ooze

cilindro cylinder

cima peak, summit

cinc *m* zinc

cinturón *m* belt

circón *m* zircon

cisterna cistern, reservoir

cizalla shear

cizallante shearing

cizallar shear

clasificar sort

clástico clastic

clasto clast

clima climate

clinómetro clinometer

clorato chlorate

clorita chlorite

cloro chlorine

cloruro chloride

cobalto cobalt

cobre *m* copper

cohesión cohesion

cohesivo cohesive

cola de pescado isinglass

colada lava flow, coulee

colgajo roof pendant

colgante suspended

colgar hang

colina hill
coloidal colloidal
coloide *m* colloid
color *m* color
columna column
columna geológica geologic column
columnar columnar
columnario columnar
coluvio colluvium
comba bulge
cometa *m* comet
compactación compaction
compactar compact
compacto compact, tight
compas *m* compass
competencia competence
competente competent
complejo complex
complementario complementary
compresa de hielo icepack
compresión compression
compresivo compressive
comprimir compress

compuesto *m* or *adj* composite, compound
compuesto acuoso slurry
cóncavo concave
concéntrico concentric
concha shell
concoideo conchoidal
concordancia conformity
concordante concordant
concreción concretion
condensación condensation
condensar condense
condensativo condensate
condrita chondrite
condrula chondrule
conducción conduction
conductibilidad conductivity
conductivo conductive
conducto conduit
congelación glaciation
congelar freeze
onglomerado conglomerate
conmoción concussion, shock
cono cone
cono aluvial alluvial fan

cono de cenizas cinder cone
cono de desmoronar scree
conodonte *m* conodont
consecuente consequent
consolidación consolidation
consolidado consolidated
contar count
continente *m* continent
contornear contour
controlar control
control estuctural
  structural control
contorno contour
convergencia convergence
convexo convex
coordenada coordinate
copo flake
coprolito coprolite
coque *m* coke
coral *m* coral
coralino coral
cordillera cordillera,
  mountain range
corindón corundum
cornalina pink agate
corneana hornfels

cornisa cornice, overhang
cornubianita hornfels
corrasión corrasion
correntera rapids
corrido thrusted
corriente *f* current, running
corriente abisal abyssal
  current
corriente de barro mudflow
corriente del lodo mudflow
corriente de resaca rip
  current
corriente de retorno *f* rip
  current
corriente en chorro *f*
  jetstream
corriente meándrica
  meandering stream
corrimiento de tierra
  landslide
corrosivo corrosive
cortar cut
cortada por las olas wave
  cut
corte *m* cut
corte delgado thin section

corteza crust
cortezal crustal
cortical crustal
costa coast
costa afuera offshore
costa posterior backshore
costero coastal, longshore
coulee *m* coulee
cráter *m* crater
cráter de impacto *m*
  impact crater
cratón craton
cresta crest, ridge
cresta afilada hogback
cresta de playa beach ridge
reta chalk
Cretáceo Cretaceous
crinoideos crinoids
criología cryology
crisoberilio chrysoberyl
crisol *m* crucible
cristal crystal
cristalino crystalline
cristalografía crystallography
cronología chronology
cuaderno notebook

cuadrado square
cuadrícula grid
cuarcífero quartz bearing
cuarcita quartzite
cuarzo quartz
cuarzo pulverisado silex
cuarzo rosado rose quartz
cuarzoso quartzose
Cuaternario Quaternary
cubierta overburden
cubierta de roca cap rock
cubierto covered
cubo cube
cuchilla hogback, razorback
cuchillo knife
cuello neck
cuenca basin
cuenca de alabeo
  downwarp basin
cuenca de drenaje *m*
  drainage basin
cuenca hidrográfica
  hydrogrphic basin
cuenca y cordillerra basin
  and range
cuerpo igneo igneous body

64

cuesta slope, hill, grade
cuesta abajo downhill
cuesta arriba uphill
cueva cave
culebra snake
cumbre *f* summit, pinnacle, top, peak
cuneiforme wedge-shaped
cuña wedge
cuprita cuprite
curso de agua watercourse
curva curve
curva de nivel contour
cúspide *f* cusp

— CH —

chaflán *m* bevel
charnela hinge
chimenea chimney, pipe, vent
chimenea hidrotermal hydrothermal vent
chimenea volcánica volcanic vent
choque shock
chorillo brook

— D —

dacita dacite
Darwiniano Darwinian
Darwinismo Darwinism
datación isotópica radiometric (isotopic) dating
datar date
dato datum
datos *mp* data
datum *m* datum
Datum Norteamericano de 1927 North American Datum 1927 NAD27
debajo de under
débil weak
decaer decay
decaimiento radioactivo radioactive decay
declinación declination
declive *m* declivity, inclination, slope
deformación deformation, strain
delantero frontal
delgado thin, lean
delizamiento landslide

delizamiento de gran alance long runout landslide

delta delta

dendrítico dendritic

dendrocronología dendrochronology

densidad density

densidad de enpaque *m* packing density

dentado jagged

deposición deposition

deposicional depositional

depósito deposit

depósito hidrotermal hydrothermal deposit

depósito de placer place deposit

depósito de yeso gypsum deposit

depresión depression, trough

depresión carsística sink

derecho straight

deriva drift

deriva continental continental drift

deriva glacial glacial drift

deriva litoral littoral drift

deriva polar polar wandering

derivar drift

derrame *m* outpouring, overflow

derretir melt

derrubio scree

derrumbamiento cave-in, collapse

derrumbamiento de rocas rockfall

derrumbamiento de tierra landslide

derrumbarse collapse

derrumbe *m* fall, collapse, slump

desagüe *m* drainage, outflow, outlet

desecar desiccate

desembocadura outlet

desembocadura de rio river mouth

desfiladero gap, gorge

desgastar corrode

desgaste attrition, mass wasting

desierto desert

deslava outwash

desliz *m* slide, slip

deslizado hacia abajo downthrown

deslizamiento slide, slip

deslizamiento de detritos debris slide

deslizamiento de gran alcance long runout landslide

deslizamiento de tierra landslide

desmenuzable friable

desmonte *m* cut

desmoronar erode

despeñadero cliff, precipice

desplazamiento creep, displacement

desplazamiento vertical throw

desplome *m* collapse

despuntado blunt

desviación deviation, diversion

detrito debris, detritus

detrito glaciario glacial outwash

Devónico Devonian

diabasa diabase

diaclasa joint

diaclasa columnar columnar jointing

diaclasado jointed

diaclasamiento jointing

diagénesis *f* diagenesis

diagonal *f* diagonal

diamante d *m* iamond

diámetro diameter

diapiro *m* diapir

diapírico diapiric

diastrófico diastrophic

diastrofismo diastrophism

diatomáceo diatomaceous

diatomea diatoms

dibujo drawing, sketch

diferenciado differentiated

diferencial differential

difícil difficult

digital digital
dilatación dilation
dilatación de congelación
  frost heave
diluir dilute
diluvio deluge, flood
diluvión *m* diluvium
dinámica dynamics
dinámico dynamic
dinosaurio dinosaur
diópsido diopside
diorita diorite
dióxido de carbono carbon
  dioxide
dique dike, levee
dirección trend
discontinuidad
  discontinuity
discordancia discordance,
  unconformity
disección dissection
diseñar draw, design
dislocación dislocation
divisoria divide, watershed
doblar fold
dolomía dolomite (roca)

dolomita dolomite (mineral)
domo dome
domo salino salt dome
domo volcánico volcanic
  dome
dorsal *f* or *adj* dorsal, ridge
dorsal centro-oceánica *f*
  mid-ocean ridge
drenaje *m* drainage
drumlin *m* drumlin
drusa geode
duna dune
dunita dunite
dureza hardness
durmiente dormant
duro hard

— E —

eclogita eclogite
ecuador *m* equator
ecuatorial equatorial
edad age
efecto Coriolis Coriolis effect
efímero ephemeral
eflorescencia efflorescence
eflorescente efflorescent

efusión effusion, outflow

eje *m* axis

ejemplar specimen

ejes *mp* axes

elástico elastic

elevación elevation, heave

elevación continental continental rise

elipsoide *m* ellipsoid

elutriación elutriation

embotado blunt

emergente emergent

emerger emerge

empinado steep

empotrar embed

empujar push

empuje *m* **ascendente** upthrust

en bruto in bulk

endógeno endogenous

energía energy

enmohecerse rust

enorme enormous, huge

enriquecimiento enrichment

ensenada cove, embayment, inlet

enterrar bury

entierro burial

Eoceno Eeocene

eólico eolian

eón *m* eon

epeirogenía epeirogeny

epeirogénico epeirogenic

epicentro epicenter

epigénesis *m* epigenesis

epigénetico epigenetic

epigénico epigene

episódico episodic

episodio episode

época epoch

época glacial ice age

epsomita Epsom salt, epsomite

era era

erizos del mar sea urchins

erosión erosion

erosionado de ola wave-cut

erosionar erode

errático *m* or *adj* erratic

erupción eruption

erupción volcánica volcanic eruption

escala scale
escala de Mohs Mohs scale
escala de tiempo time scale
escalón *m* echelon
escama flake
escamoso flaky, scaly
escape *m* outlet
escarcha frost
escarpa escarpment, scarp
escarpado steep
escollo reef
escombrera dump
escombro de talud scree
escombros debris, rubble
escoria cinder, clinker, scoria, slag
escudo shield
escudo continental continental shield
esfera sphere
esférico spherical
esferoide *m* spheroid
esker *m* esker
esmeralda emerald
esmeril *m* emery
espacio interval

esparcir spread
espato spar
espato de Islandia Iceland spar
especie *m* species
espejo de fricción slickensides
espeleología speleology
espeso dense, thick
espesor *m* thickness
espinazo hogback
espinela spinel
espiral *f* or *adj* spiral
espiral de Ekman *f* Ekman spiral
espodumeno spodumene
esqueleto skeleton
esquina corner
esquisto schist
esquistoso slaty
estable stable *adj*
estación season
estalactita stalagtite
estalagmita stalagmite
estaño tin
estanque *m* pond

este east
estero estuary
estratificación bedding, stratification
estratificación cruzada cross bedding
estratificación transversal cross bedding
estratificado layered, stratified
estratigrafía stratigraphy
estratigráfico stratigraphic
estrato layer, stratum
estratos strata
estratosfera stratosphere
estratovolcán stratovolcano
estrecho narrow, straits
estrella star
estrés *m* stress
estresar stress
estría *fs* stria
estriación striation
estriar striate
estrías *fp* striae
estribación foothill, spur
estribo abutment, spur

estromatolita stromatolite
estroncio strontium
estructura structure
estuario estuary
etapa stage
evaporación evaporation
evaporar evaporate
evaporita evaporite
evapotranspiración evapotranspiration
evolución evolution
excavación para minar stope
excavar excavate
excavar hacia abajo downcut
exfoliación exfoliation
exfoliar exfoliate
exhumación exhumation
expansión expansion
expansión de los fondos oceánicos seafloor spreading
expeler eject, extrude
experimental experimental
explosión explosion

exposición exposure
exsolución exsolution
extensión extention, sheet
exterior *m* or *adj* exterior
extraer pluck
extrusión extrusion
extrusivo extrusive
eyectar eject
eyecto ejectum
eyestos ejecta

— F —

faceta facet
facies *fs* facies
faja strip, zone
fajas de alisios trade winds zones
falla fault
falla a la derecha dextral fault
falla de cabalgadura thrust fault
falla de cizalla shear fault
falla de corrimiento thrust fault

falla de desgarre detachment fault
falla de desplazamiento de rumbo strike-slip fault
falla de desprendimiento detachment fault
falla de inclinación dip fault
falla de relajación release fault
falla de rumbo strike-slip fault
falla dextral dextral fault
fallado faulted
falla escolonada en echelon fault
falla inversa thrust fault
falla lateral derecha right-lateral fault
falla lateral izquierda left-lateral fault
fallamiento faulting
falla normal normal fault
falla pivotal scissors fault
falla ramificada branching fault

falla transformante transform fault

fanerítico coarse-grained (phaneritic)

fanglomerado fanglomerate

fango mud, ooze

farallón *m* bluff, sea stack

fase phase

fauna fauna

faz facies

fechamiento isotópica (radiométrico) radiometric (isotopic) dating

fecha date

fechar to date

feldespático feldspathic

feldespato feldspar

félsico felsic

fenoclástico phenoclastic

fenoclasto phenoclast

fenocristal phenocryst

férrico ferric

ferrito ferrite

ferrocarril railway

ferromagnesiano ferromagnesian, mafic

ferroso ferrous

figura de relieve boss

filón lode, vein

filtración percolation, seepage

filtrarse seep, ooze

filtro filter

física physics

físico physical

fisíl fissile

fisiográfico physiographic

fisura fissure

flaco thin, skinny

flanco flank

flora flora

fluido fluid

flujo *m* or *adj* flow

flujo de lodo mudflow

flujo y reflujo ebb and flow

flúor *m* fluorine

fluorescencia fluorescence

fluorescente fluorescent

fluorita fluorite

fluoruro fluoride

fluvial fluvial

fluvoglacial fluvoglacial

foco focus
foliación foliation
foliar foliate
fondo bottom, floor
foraminíferos foraminifera
forma fisográfica landform
forma de relieve landform
formación formation
fosa ditch, trench
fosa oceánica submarine trench
fosa submarina submarine trench
fosa tectónica graben
fosfato phosphate
fosforita phosphorite
fósil *m* or *adj* fossil
fósil índice index fossil
fosilífero fossiliferous
fosilización fossilization
fosilizar fossilize
foso pit, hole
fotografía photograph
fracaso failure
fraccionario fractional
fractura fracture

fractura por congelamiento frost heave
fragmento fragment
franja fringe
freático phreatic
freatofito phreatophyte
frente *f* face, front
frente, de frente frontal
friable friable
fricción friction
fuego fire
fuente *f* fountain, source, spring
fuerte strong
fuerza force
fuerza gravitatoria gravitational force
fulgurita fulgurite
fumarola fumarole
fumarola negra black smoker
fundamental fundamental
fundido molten
fundir melt

— G —

gabro gabbro

galena galena
galería gallery
galería de excavación adit
ganga gangue
garganta gorge, ravine, water gap
gas gas
gas natural natural gas
géiser *m* geyser
gema gem
gemelo twin
genético genetic
geocronología geochronology
geoda geode
geode *f* vug
geodesía geodesy
geodésico geodesic
geodético geodetic
geofísica geophysics
geografía geography
geográfico geographic
geoide *m* geoid
geología geology
geología de prospección exploration geology
geología estructural structural geology
geología histórica historical geology
geológicamente geologically
geólogo geologist
geomagnético geomagnetic
geomorfología geomorphology
geoquímica geochemistry
geosinclinal *m* geosyncline
geotérmico geothermal
giba hump
giro gyre
glaciación glaciation
glacial glacial
glaciar glacier
glaciario glacial
glacifluvial fluvioglacial
globigerina globigerina
globo globe
globular globular
gneis *m* gneiss
golfo gulf
Gondwana Gondwana
graben graben

gradiente *f* gradient
grado degree
grafito graphite
granate garnet
grande large
granítico granitic
granito granite
granizo hail
grano grain
granodiorita granodiorite
granular granular, granulate
granularidad granularity
gránulo granule
graptolites *mp* graptolites
grava gravel
grava de playa beach gravel,
   shingle
gravedad gravity
gravilla grit
gravitación gravitation
greda marl
grieta crack, crevasse,
   crevice, fissure
gris gray
grit grit
grueso coarse

grus *m* grus
gruta grotto
gubia gouge
guija pebble
guijarro cobble
guijo shingle
guijón *m* cobble
guyot *m* guyot

— H —

hacer añicos shatter
hacia la orilla inshore
halita halite
halo halo
halogéno halogen
helada frost
helamiento glaciation
helar freeze
helero glacier
hematita hematite
hematita mena de hierro
   hematite iron ore
hematites *fs* hematite
hendible fissle
hendidura rift
herrumbre *f* rust

herrumbroso rusty
hervidero boiling spring
hervir boil
hiato hiatus
híbrido hybrid
hidratación hydration
hidratar hydrate
hidrato *m* hydrate
hidráulico hydraulic
hidrita hydrite
hidrocarburo hydrocarbon
hidrología hydrology
hidrostático hydrostatic
hidrotermal hydrothermal
hielo ice
hielo pastoso sludge ice
hierro iron
hipsométrico hypsometric
hoja de papel sheet
Holoceno Holocene
homoclinal homocline
homogéneo homogeneous
hondo deep
hondonada gully
hora hour
horizonal horizontal

horizonte *m* horizon
hornablenda hornblende
hornfelsa hornfels
horno oven (hornito)
horst *f* horst
hoya basin, hole, trough
hoyo pit, hole
hoyo glaciario kettle
hoyoso pitted
hueco *m* or *adj* hollow
huella footprint, track
huesa grave
hueso bone
hulla bituminous coal
húmedad moisture
húmedo damp
húmico humic
humus *m* humus
hundimiento subsidence
hundirse founder, sink,
    subside, collapse
huracán *m* hurricane

— I —
iceberg iceberg
ígneo igneous

77

ignimbrita ignimbrite, welded pumice, welded tuff
ilmenita ilmenite
iluviación illuviation
iluvial illuvial
iluvio illuvium
imán *m* magnet
impactita impactite
impermeable impermeable
inciso incised
inclinación dip, inclination, slope, grade, pitch
inclinación abajo downdip
inclinar incline, dip
inclinarse tilt
inclusión inclusion
incompetente incompetent
índice index
induración induration
indurar indurate
inestable unstable
infiltrar percolate
informe report
ingeniero engineer
inicial initial
insecto insect

inselberg *f* inselberg
insolación insolation
insoluble insoluble
intensidad intensity
intercalación intercalation
intercalar intercalate
interestatal interstate
interestratificado interbedded
interetapa interstage
interglacial interglacial
inerior *m* or *adj* interior
interlobulado interlobate
intermedio intermediate
intermitente intermittent
intermontano intermontane
interno intern, internal
intersticial intersticial
intersticio interstice
intervalo interval
intricado intricate
intrusión intrusion
intruso intrusive
inundación flood, inundation
inundar drown, flood, inundate
invasión invasion

invasor invasive
inversión inversion
inverso inverse
invertido inverted,
   overturned
invertir invert
inyección injection
inyectar inject
iridio iridium
isla island
isoclinal isocline
isócrona isochrone
isomorfo isomorph
isomórfico isomorphic
isopaca isopach
isostasia isostacy
isostático isostatic
isotópico isotopic
isótopo isotope
isotropia isotropy
isotrópico isotropic
istmo isthmus

— J —

jaboncillo gouge
jade jade

jadeita jadeite
jaspe jasper
joven young
juntura joint
jungla jungle
Jurásico Jurassic

— K —

kame *m* kame
kimberlita kimberlite

— L —

labrador *m* labradorita (rock)
labradorita labradorite
   (mineral)
lacolito laccolith
lacustre lacustrine
ladearse tilt
lado side, lateral
lado caido downthrown side
lado levantado upthrown side
lago lake
lago seco dry lake
laguna lagoon
laguna de desplome sag
   pond

laguna semilunar oxbow lake

lahar *m* lahar

laja slab, flagstone

lama mud, ooze, slime

lámina lamina, sheet

lámina delgada thin section

laminado laminated, platy

laminella lamella

lapis lázuli *m* lapis lazuli

lápiz pencil

largo long

latente dormant

lateral *adj* lateral, side

laterita laterite

laterítico lateritic

latitud *f* latitude

latón *m* brass

Laurasia Laurasia

lava lava

lava cordada pahoehoe

lava fluida pahoehoe

lavaje *m* **de pendiente** slope wash

lavar wash

lava (vicosa) en bloques aa

lechada slurry

lecho base, bed, layer, riverbed

lecho de la roca bedrock

lechos strata

légamo slough

lengua de tierra spit

lente *f* lens

lente *f* **de aumento** magnifying glass

levantado thrown

levantamiento uplift, rise, elevation, heave

levantar raise, lift, elevate

levee *f* levee

liga alloy

ligero light

lignito lignite

lijar con arena sandblast

lima lime

limo mud, silt, slime

limonita limonite

limoso silty

limpiar scour (channel)

lindero abutment

línea costera coast line

línea de base baseline

linéa de cataratas por una meseta fall line
línea de falla fault line
lineal lineal, linear
lineamento lineament
linear linear
líquido *m* or *adj* liquid
liso smooth
lista band
lístrico listric
lítico lithic
litificación lithification
litificado lithified
litio lithium
litografía lithography
litográfico lithografic
litología lithology
litoral *m* or *adj* littoral
litosfera lithosphere
litostático lithostatic
lixiar leach
lobulado lobate
lóbulo lobe
lodo mud
lodo de perforación drilling mud

lodoso muddy
lodo volcánico volcanic mud
loess *m* loess
loma hill
longitud longitude, length
longitud de onda wavelength
losa flagstone, slab
loxodromía loxodrome
luna moon
lupa hand lens, magnifying glass
lutita shale
lutítico shaly
lutolita mudstone
luz *f* light

— LL —

llamada de fondo upwelling
llano flat, plain
llanura plain
llanura abisal abyssal plain
llanura aluvial alluvial plain
llanura costanera coastal plain
llover rain
lluvia rain

81

lluvia de cenizas ash fall, ash shower

— M —

maar maar
macizo *m* massif
macizo massive
machete knife
madurez *f* maturity
maduro mature
máfico mafic
magma magma
magmático magmatic
magnesia magnesia
magnesiano magnesian
magnesio magnesium
magnesita magnesite
magnetismo magnetism
magnetita magnetite, lodestone
malaquita malachite
malpaís *m* badlands
mamelón mamelon
mamut *m* mammoth
manantial *m* spring

manantial de agua termal *m* hot spring
manantial caliente *m* hot spring
manantial herviente *m* boiling spring
mancha stain
manga waterspout
manganeso manganese
manto mantle, sheet
manto de hielo ice sheet
mapa *m* map
mapa geologico geologic map
mar *m* or *f* sea
mar adentro offshore
mar antiguo relic sea
marca mark
marcación bearing
marcar mark
marea tide
marea alta high tide
marea muerta neap tide
maremoto submarine earthquake
marga loam, marl

margen margin, border, edge, fringe

margen *m* **continental** continental margin

marginal marginal

marino marine

marisma mudflat

mármol *m* marble

mar playo shallow sea

martillo hammer

más alto upper

más bajo lower

más reciente late

masa mass

masa infrayacente undermass

masa suprayacente overmass

masa terrestre landmass

materia matter

matrix matrix

meandro meander

meandro abandonado oxbow lake

meandro cortado intrenched meander

meandro excavado incised meander

mecánica de suelos soil mechanics

mecánico mechanical

mecánica mechanics

médano dune

media vida half-life

medial medial

mediana median

mediano medium

medio half, middle

medio ambiente environment

mediterráneo mediterranean

mellado jagged

mena ore

mena de hierro iron ore

mena negra black ore

mercurio mercury

meridiano meridian

meridional meridional, southern

mesa mesa

meseta plateau

meseta tectónico horst

mesosfera mesosphere

Mesozoico Mesozoic

metal *m* metal

metamórfico metamorphic

metamorfismo metamorphism

metamorfismo regional regional metamorphism

metano methane

meteórico meteoritic, meteor, meteorite

meteorización weathering

meteorizado weathered

meteorizar weather

meteoro meteor

meteorología meteorology

mica mica

micáceo micaceous

micacita micaschist

microbrecha microbreccia

microclina microcline

microcristalino microcrystaline

microscopio microscope

microfósil *m* microfossil

miembro member

migración de los polos wandering of the poles

milonita mylonite

milonítico mylonitic

milonitización mylonitization

mina mine

mina a cielo abierto open pit mine, strip mine

mineral *m* or *adj* mineral

mineralización mineralization

mineralizar mineralize

mineralogía minerology

minería mining

minuto minute

mioceno miocene

Mississippiense Mississippian

Misuriense Missourian

mitad half

mixto mixed

mojo rust

mojón bench mark

molde *m* c ast

molécula molecule

moler grind, mill

molibdeno molybdenum

molusco mollusc

monacita monazite

monadnok monadnock

monoclinal *m* monocline

montaña mountain
montaña isla inselberg
montañas highland
montañoso mountainous
monte mount
montecillo hummock, hump
monte submarino *m* seamount
montículo hummock, mound
montmorillonita montmorillonite
monzón *m* monsoon
monzonita monzonite
moreno brown
morrena moraine
morrena de retroceso recessional morraine, retrocessional morrainem
morrena intermedial medial moraine
morrena lateral lateral moraine
morrena lobulada lobate moraine

morrena terminal terminal morraine
movido hacia abajo downthrown
muestra sample, specimen
muestra de suelo soil sample
muestra de testigo core sample
muestrear sample
mundial global
muralla wall (rock)
muro wall (rock)
muro colgante hanging wall
muro de base footwall
muscovita muscovite

— N —

nacimiento birth
naríz *f* nose
nativo native
natural natural
naturaleza nature
náutico nautical
nautiloideos nautiloids
neblina mist
necton *m* nekton

nefelina nepheline
nefelinita nephelinite
nefrita nephrite
negativo negative
negro black
neis *m* gneiss
Neogeno Neogene
nevar snow
nevasca blizzard, snow storm
niebla fog
nieve *f* snow
níquel *m* nickel
nitrato nitrate
nitrógeno nitrogen
nivación nivation
nivel *m* level
nivel básico base level
nivel de base base level
nivel de mano hand level
nivel del mar sea level
nivel de referencia
   reference level
nivel freático water table
no fijo unattached
no perturbado undisturbed
nodo node

nódulo nodule
nódulo de manganeso
   manganese nodule
normal normal
normas standards
norte *m* or *adj* north
nube *f* cloud
nube ardiente *f* nuee
   ardiente
núcleo core
núcleo de pozo drill core
núcleo interior inner core
núcleo metálico metalic core
nudo knot
nunatak *m* nunatak

— O —

oasis *m* oasis
oblado oblate
oblicuo oblique
obsidiana obsidian
obturado plugged, sealed
obturador plug
obtuso obtuse
oceánico oceanic
océano ocean

ocilación swing
ocre *m* ocher
oeste west
ofiolita ophiolite
ola wave
ola de marea tidal bore
oligoceno oligocene
oligoclasa oligoclase
olivino olivine
onda wave
onda de choque shock wave
onda P P wave
onda S S wave
onda secundaria secondary wave
onda sísmica seismic wave
onda sísmica marina seismic sea wave
ónix *m* onyx
oolito oolite, oolith
opaco opaque
ópalo opal
órbita orbit
Ordovícico Ordovician
orientación bearing
orientar orient

origen origin
orilla bank, shore, fringe
orín *m* rust
orla cusp
oro gold
oro de tontos fools' gold
orogénesis *m* orogeny
orogenia orogeny
orogénico orogenic
orogénio orogene
orográfico orographic
ortoclasa orthoclase
oscilación swing
oscilar ocillate
oscuro dark
osmio osmium
otero knoll
oxidación oxidation
oxidar oxidize
óxido oxide
oxígeno oxygen
ozono ozone

— P —

pahoehoe pahoehoe
paisaje landscape

paladio paladium
Paleoceno Paleocene
Paleogéno Paleogene
paleontología paleontology
paleosol *m* paleosol
Paleozoico Paleozoic
palinespastico palinspastic
pampa pampa
pan de azúcar sugarloaf
pando bulged
Pangea Pangea
pantano bog, swamp, marsh
pantanoso swampy
paquete packet
par estereoscópico stereo pair
paralelo parallel
parálico paralic
pared wall
partícula particle
pasaje passage, straits
pasillo corridor
paso pass, pace
pasta aguada slurry
pavimento pavement
pavimiento del desierto
    desert pavement

pechblenda pitchblende
pedacito chip
pedazo piece, shard
pedernal *m* flint
pedestal *m* butte
pedimento pediment
pedología pedology
pedregoso rocky
pedrejón *m* boulder
pedrero scree
pegajoso sticky
pegmatita pegmatite
pelágico pelagic
Peleano Pelean
pelotilla pellet
pendiente *f* gradient, slope
pendiente de deslizamiento
    slip-off slope
pendiente *f* **de escombros**
    talus slope
pendiente socavada
    undercut slope
peneplanicie peneplain
peneplanicie encañada
    dissected peneplain
peninsula peninsula

Pennsilvanico Pennsylvanian

peña cliff, crag, rock

peñasco cliff, bluff, large rock

peñascoso rocky

pepita nugget

percolación percolation

percolar percolate

perfador driller

perfecto perfect

perfil *m* profile

perfil geológico geologic profile

perforar drill, bore, perforate

pergelisol *m* permafrost

peridotita peridotite

peridoto peridot

período period

perlita perlite

permeable permeable, pervious

Pérmico Permian

perspectiva perspective

perturbación disturbance

pesado heavy

peso load, weight

pétreo rocky, stony

petrificado petrified, lithified

petrificar petrify

petróleo petroleum, oil

petrología petrology

petrolutita oilshale

piamonte *m* piedmont

pico peak, pinnacle

pico solitario nunatak

pico submarino seamount

pie de montaña foothill

piedra stone

piedra de afilar sharpening stone

piedra de chispa flint

piedra imán lodestone

pieza piece

pilar *m* pillar

pilar de roca *m* hoodoo

pirita pyrite

piroclástico pyroclastic

piróxeno pyroxene

piso bottom, floor

pista trail

pizarra slate
pizarroso slaty
placa plate
placa tectónica tectonic plate
placer placer
placer deposito placer deposit
placer eluvial *m* dry diggings
plagioclasa plagioclase
plagioclasis plagioclase
planada level ground
plancha slab
plancheta plane, table
plancton *m* plankton
planeta planet
planetesmal *m* planetismal
planicie *f* plain
planicie costanera coastal plain
plancie de marea tidal flat
plano *m* or *adj* flat, plane
plano de estratificacion bedding plane
plano de falla fault plane
planta plant
plasticidad plasticity

plata silver
plataforma shelf, platform
plataforma continental continental shelf, continental platform
plataforma Precámbrica Precambrian platform
platino platinum
playa beach, dry lake, playa
playa de la marea foreshore
playa efímera playa lake
playa enbarrera barrier beach
playa posterior back beach
playón large shore
pleamar flood tide, hightide
plegado folded
plegamiento folding
plegar fold
Pleistoceno Pleistocene
pliegue *m* fold
pliegue anticlinal *m* anticlinal fold
pliegue en «V»
pliegue *m* **reclinado** recumbent fold

pliegue *m* **recostado** recumbent fold

pliegue *m* **sinclinal** synclinal fold

Plioceno Pliocene

plomo lead

pluma pen

plutón pluton

plutónico plutonic

pluvial pluvial

poco profundo shallow, shoal

polar polar

polen *m* pollen

policónico polyconic

pólipo polyp

polo pole

polo migratorio wandering pole

polvo dust, powder

pómez *m or f* pumice

pongo gorge

pórfido porphyry

porfirítico porphyrtic

pororoca tidal force

porosidad porosity

poroso porous

posglacial postglacial

posición attitude

positivo positive

poste post

posterior subsequent

postura attitude

potasa potash

potasio potassium

poza pool

pozo well

pradera prairie

prado meadow

Precámbrico Precambrian

precipicio cliff, precipice

presa dam

preservación preservation

presión pressure

primario primary

principio principle

profundidad depth

profundidad oceánica deep-sea

profundizar deepen, incise

profundo *adj* deep

promedio average, mean

promontorio headland, promontory
proporción ratio
prospectar prospect
Proterozoico Proterozoic
protuberancia boss, knob
provincia province
proximidad proximity
proyección projection
proyección UTM UTM projection
pseudomorfo pseudomorph
puente *m* bridge
puente *m* **natural** natural bridge
pulverisar pulverize
pumicita pumice
pumicita soldada welded pumice
punta point (of), cape
punto point (to)
punto caliente hot spot
punto topográfico benchmark

— Q —

quebrada gulch, ravine
quebrado broken
quebrar break
química chemical
química chemistry
químico chemist

— R —

rada bight
radio radius
radioactividad radioactivity
radiolarios radiolaria
radón *m* radon
raedura abrasion
raer abrade
rajadizo fissle
rajadura crack
rajadura a cielo abierto open cut
rama branch
ramal *m* branch
ramificar branch
rango rank
rapidez rapidity, speed
rápidos rapids

rasgo abierto open cut, open trench

rasgo estructural structural feature

rasguño scratch

raspadura strëak

rastro trace

rayadura streak

rayo lightning flash, ray

rayo ahorquillado forked lightning

razón *m* rate

reactivación rejuvenation

rebajamiento planation

rebanada slice

rebote *m* rebound

rebote elástico elastic rebound

recarga de agua subterránea ground water recharge

receptáculo resevoir

reciente recent

recodo bend

recolectar collect, sample

reconocer reconnoiter

reconocimiento reconnaissance

recorrer traverse

recorrido route

rectángulo rectangle

recuperación recovery

recuperado recovered

recursos minerales mineral resources

red lattice, mesh, net

redimiento output, yeild

redondeamiento rounding

redondez *f* roundness

redondo rounded

reemplazo replacement

reflección sísmica seismic reflection

reflexión reflection

reflujo ebb

refracción refraction

refracción sísmica seismic refraction

regir control

registro record

registro geológico geologic record

regla de cálculo slide rule

regolito regolith

regresión regression

regulación estructural structural control

rejuvenecimiento rejuvenation

relación relation, ratio, report

relámpago lightning

relaves *mp* tailings

relicto relict

relicto de erosión butte

relicto exterior outlier

relicto interior inlier

relieve *m* relief

relleno fill

remoción en masa mass wasting

remolino eddy, vortex, whirlpool, whirlwind

repasar review

replica aftershock

represa dam

reptación creep

resaca undertow

resbalamiento slip

resbaloso slippery

resección cutoff

residual residual

resina resin

resistencia strength, resistence

resistente resistent

respiradero vent

restinga rock ridge, rock beach barrier, beach ridge

retallo ledge

retardo lag

reticulado grid

retirada recession

retirar remove

retroceso recession, retreat

retroseso glacial glacial recession

reversión reversal

revisar review

rezagarse lag

rezumarse ooze, seep

riada flood

ribera river bank

ribera posterior backbeach, backshore

ribero levee
rincón interior corner
río river
río abajo downstream
río arriba upstream
riolita rhyolite
ripio ballast, rubble
risco crag, cliff, bluff
ritmo rhythm
rizadura ripple
roca rock
roca aborregada roche moutonee
roca afanítica aphanitic rock
roca apartada outlier
roca arcillosa claystone, mudstone
roca arenisca sandstone
roca cubierta cap rock
roca de fondo bedrock
roca de lecho bedrock
roca decubierta denudate rock
roca encaramada perched rock

roca eruptiva eruptive rock, extrusive rock
roca extrusiva extrusive rock
roca fanerítica phaneritic rock
roca firme bedrock
roca grande large rock
roca ignea igneous rock
roca intrusiva intrusive rock
rocalla debris
roca madre country rock
roca metamórfica metamorphic rock
roca metamórfica de contacto contact metamorphic rock
roca monegote hoodoo
roca sedimentaria sedimentary rock
roca separado perched rock
roca sólida bedrock
rocoso rocky
rodocrosita rhodochrosite
rojo red
rómbico rhombic
rompeolas breakwater
romper break

rompiente *m* breaker, surf

rosado pink

roto broken

rotura breaking

rubí *m* ruby

rubidio rubidium

rugoso wrinkled

rumbo trend, bearing, course, direction

rumbo de la falla strike of fault

rumbo del estrato strike

ruptura rupture

ruta route

rutilo rutile

— S —

sabana savanna

sábana sheet

sacudida shake

sacudir shake

sal *f* salt

salado salty

sal *f* **de Epsom** Epsom salt, espomite

sal gema salt crystal

salida outlet

salina salina, salt flat, salt mine

salino saline

salir emerge

salitre saltpeter

salmuera brine

salobre brackish

salto de agua waterfall

sangre blood

satélite *m* satellite

sección section

sección delgada thin section

sección transversal cross-section

secuencia sequence

sedimentación sedimentation

sedimentario sedimentary

sedimento deposit, sediment

sedimentología sedimentology

seguro safe, secure

seleccíon sorting

selecciónar sort

selección natural natural selection

selenio selenium

selenita selenite

selva jungle, forest, esp. rain forest

semiárida semiarid

senda path, footpath, trail

sendero path, footpath, way

senectud *f* senility

senil old

seno bight, sine, trough

sentado sessile

separación offset

septentrional northern

sepultar burry

sequía drought

serie series, serial

serpentina serpentine

serpiente *f* serpent, snake

serración serration

Servicio Geológico Geological Survey

sésil sessile

seudomorfo pseudomorph

sial sial

sidéreo sidereal

siderita siderite

sienita syenite

sierra range, sierra

sílex chert

silicificación silification

silicato silicate

sílice silica

silíceo siliceous

silicio silicon

silo sill

silla saddle

Silúrico Silurian

sima abyss, chasm, sima

símbolo symbol

simetría symmetry

simétrico symmetric

similar similar

sinclinal *m* syncline

sísmico seismic

sísmo earthquake

sismógrafo seismograph

sismología seismology

sismólogo seismologist

sismómetro seismometer

sistema system

sistema global de posición Global Positioning System GPS

sitio, en sitio in situ, in place

skarn *m* skarn

sobrecapa overburden

sobrecarga overload

sobrecolgante overhanging

sobrecorrido overthrust

sobrecorrimiento overthrust

sobreempujado overthrusted

sobreimponer superimpose

sobrelevantado upthrown

sobrencia seepage

sobresalir overhang

socavación sapping, undercutting

socavado undercut

socavar undercut

soda soda

sodio sodium

sol sun

solapamiento overlapping

solapar overlap

solevantamiento upthrust

sólido *m or adj* solid

soluble soluble

solución solution

soluto solute

somero shallow

sondeo sounding

soplar blow

sotavento *m or adj* lee

sturzstrom sturzstrom, long-runout landslide

suave smooth, soft

subducción subduction

subducir subduct

submarino *adj* submarine

subsecuente subsequent

subsuelo subsurface, subsoil

subyacente underlying

sucesión sequence

suelo ground, soil

suelto loose

sulfato sulfate

sulfuroso sulphurous

sumario summary

sumidero sink hole

superficie *f* surface

superficie *f* **de deslizamiento** slickensides

supracortical supercrustal

suprayacente overlying

suprayacer overlie

sur south

surcado furrowed, grooved

surco groove

surgencia upwelling

surgente upwelling

surgir emerge

suspender suspend

suspensión suspension

— T —

tabla slab

tajo abierto open cut

taladrar bore, drill

taladro *n* drill

talco talc

talud *m* slope, talus

talud continental continental slope

tamaño de grano grain size

tamiz *m* sieve

tangente tangent

tapón *m* plug

tapón volcánico volcanic plug

tarído late

taxonomía taxonomy

tectita tektite

tectónica de placas plate tectonics

tectónico tectonic

tefra tephra

teja shingle

telescopio telescope

temblor *m* earthquake, temblor, tremor

temblor previo foreshock

témpano floe

témpano de hielo ice floe, iceberg

temprano early

temperatura temperature

tensión tension

tensionar tension

teodolito theodolite

Terciario Tertiary

termoclina thermocline

termosfera thermosphere

terraplén *m* bank, embankment, fill, terrace

terraza terrace

terremoto earthquake

terreno terrain, terrane, soil

terreno costroso scabland

terrenos de acarreo glacial drift

terrestre terrestrial

terrígeno terrigenous

terrón *m* clod

terroso earthy

testigo core, drill core

Tetis Tethys

textura texture

tiempo time, weather

tiempo geológico geologic time

tierra dirt, earth, ground land, soil

tierra de Fuller Fuller's earth

tierra firme landmass

tierras altas highland

tierras raras rare earths

tifón typhoon

till *m* boulder clay, till

tillita tillite

tirar pull

titanio titanium

tixotrópico thixotropic

tiza chalk

toba sinter, tuff

toba soldada welded tuff

tobáceo tuffaceous

toleítico tholeitic

tómbolo tombolo, sandbar

tomografía tomography

tonalita tonalite

topacio topaz

topografía topography, terrain, terrane, surveying

topográfico topographic

topógrafo surveyor, topographer

torio thorium

tornado tornado

torrente *m* torrent

tosca caliche, hardpan

tosco coarse

transgresión transgression

transitorio intermittent

transporte transport
trap *m* trap (rock)
traquita trachyte
traslapar overlap
traslapo overlap
travertino travertine
traza trace
traza de falla fault trace
triángulo triangle
Triásico Triassic
trilobites *mp* trilobites
trinchera trench
trinchera abierta open
   trench
trinchera submarina
   submarine trench
trípode tripod
triturar crush, grind,
   pulverize, shatter, triturate
trizas shards
tromba spout
tromba marina waterspout
trocha trail
tropical tropical
trópico tropic
troposfera troposphere

trozo fragment
trueno thunder
truncar truncate
tsunami *m* tsunami
tubo pipe
tubería pipe
tufa tufa
túmulo tumulus
tundra tundra
túnel tunnel
turba peat
turbidita turbidite
turmalina tourmaline
turquesa turquoise

— U —

ultrabásico ultrabasic
ultramáfico ultramafic
unida a la costa spit
uniformitarianismo
   uniformitarianism
Universal transversa de
   Mercator UTM Universal
   Tranverse Mercator
uraninita uraninite,
   pitchblende

uranio uranium

— V —

«V», en «V» chevron
vadear ford
vado ford
valle *m* valley
vallecito dale
valle colgante *m* hanging
  valley
valle de fractura *m* rift
  valley*m*
valle en forma de «U» *m*
  U-shaped valley
valle en forma de «V» *m*
  V-shaped valley
valle inundado *m* drowned
  valley
valuar rate
valva shell
vapor *m* vapor
variación variation
varve varva
velocidad speed, velocity
vena vein
venilla veinlet, veinule

ventana window
ventifacto ventifact
ventisca blizzard
verde green
vereda path, trail
vermiculita vermicullite
vertedero spillway
vertical vertical
vértice *m* apex, vertex
vertiente *m* watershed,
  spring
vesícula vesicle
vesicular vesicular
Vesubio Vesuvius
veta band, lode, seam,
  vein
vida media half life
vidrio glass
viejo old
viento wind
vientos alisios trade winds
viscosidad viscosity
vítreo glassy, vitreous
vitriolo vitriol
volcán *m* volcano
volcán activo active volcano

volcán apogado extinct volcano

volcán compuesto composite volcano

volcán *m* **de lodo** mud volcano, hervidero

volcán *m* **en actividad** active volcano

volcán *m* **en escudo** shield volcano

volcán extinto extinct volcano

vocánico volcanic

volcán submarino undersea volcano

volframio wolfram

volumen, en volumen in bulk

vug vug

yesera gypsum deposit

yeso gypsum

zafiro sapphire

zanja ditch, trench

zanja de desagüe coulee

zanja geológica graben

zeolita zeolite

zona zone, belt

zona de baja velocidad low velocity zone

zona de falla fault zone

zona fronteriza borderland

zoolita zoolith

zoología zoology

— W X Y Z —

wacka wacke

wolram *m* wolfram

xenolito xenolith

yacimiento bed, deposit, reservoir

Charts and Tables

Cartas y Tablas

Geologic Time Scale

Era	Period	Epoch	Millions of Years Before Present
Cenozoic	Quaternary	Holocene	0-0.01
		Pleistocene	0.01-1.6
	Tertiary	Pliocene	1.6-5.0
		Miocene	5.0-24
		Oligocene	24-38
		Eocene	38-55
		Paleocene	55-66
Mesozoic	Cretaceous		66-138
	Jurassic		138-205
	Triassic		205-240
Paleozoic	Permian		240-290
	Carboniferous		
	Pennsylvanian		290-330
	Mississippian		330-360
	Devonian		360-410
	Silurian		410-435
	Ordovician		435-500
	Cambrian		500-570
Precambrian	Proterozoic		570-2,500
	Archeozoic		2,500-4,500
Origin of Earth	·		4,500

Source: USGS *STA* 7th ed., 1991, 59, based on Geologic Names Committee (1980).

CRONOLOGÍA GEOLÓGICA

Era	Período	Época	Milliones de Años Antes del Presente
Cenozoico	Cuatertario	Holoceno	0-0,01
		Pleistoceno	0,01-1,6
	Terciarío	Plioceno	1,6-5,0
		Mioceno	5,0-24
		Oligoceno	24-38
		Eoceno	38-55
		Paleoceno	55-66
Mesozoico	Cretácio		66-138
	Jurásico		138-205
	Triásico		205-240
Paleozoica	Pérmico		240-290
	Carbonifero		
	Pensilvaniense		290-330
	Mississipiense		330-360
	Devónico		360-410
	Silúrico		410-435
	Ordovicio		435-500
	Cámbrico		500-570
Pre-cámbrico	Proterozoico		570-2.500
	Arqueozoico		2.500-4.500
Orígen de la Tierra			4.500

IGNEOUS ROCK CHART

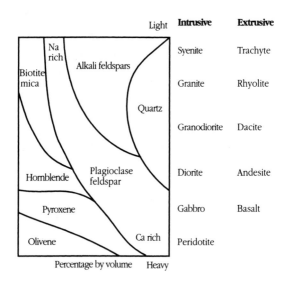

	Intrusive	Extrusive
	Syenite	Trachyte
	Granite	Rhyolite
	Granodiorite	Dacite
	Diorite	Andesite
	Gabbro	Basalt
	Peridotite	

Light

Na rich

Alkali feldspars

Biotite mica

Quartz

Hornblende

Plagioclase feldspar

Pyroxene

Olivene

Ca rich

Percentage by volume

Heavy

Source: Dieterich, R. V. 1970. *Geology and Virginia*. Richmond: University Press of Virginia.

Diagrama de Rocas Igneas

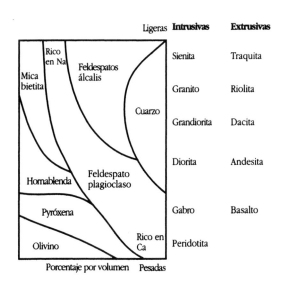

	Ligeras	**Intrusivas**	**Extrusivas**
		Sienita	Traquita
		Granito	Riolita
		Grandiorita	Dacita
		Diorita	Andesita
		Gabro	Basalto
		Peridotita	

Rico en Na

Mica bietita

Feldespatos álcalis

Cuarzo

Hornablenda

Feldespato plagioclaso

Pyróxena

Olivino

Rico en Ca

Porcentaje por volumen   Pesadas

GRAIN SIZE TABLE

Grain Diameter*	Grain-size Name	Class	Sedimentary Rock Name	
	Boulder			
256	(≈10 inches)			
	Cobble			
64	(≈2.5 inches)	Gravel	Conglomerate	
	Pebble			
4	(≈0.16 inch)			
	Granule			
2				
	Very Coarse			
1				
	Coarse			
0.5				
	Medium	Sand	Sandstone	
0.25				
	Fine			
0.125				
	Very Fine			
0.0625				
	Silt		Siltstone	(>2/3 silt)
0.0039		Mud	Mudstone	(subequal silt & clay)
	Clay		Claystone	(>2/3 clay)

* millimeters

Source: Sediment sizes based on scale developed by J. A. Udden in 1898. Names of size classes standardized by C. K. Wentworth in 1922.

Tabla con Tamaños de Granos

Diámetro de Grano*	Nombre de Tamaño de Grano	Clase	Nombre de Roca Sedimentaria	
	Pedreón			
256	(≈10 pulgadas)			
	Guijarro			
64	(≈2,5 pulgadas)	Grava	Conglomerado	
	Guija			
4	(≈0,16 pulgada)			
	Gránulo			
2				
	Muy Gruesa			
1				
	Gruesa			
0,5				
	Mediana	Arena	Arenisca	
0,25				
	Fina			
0,125				
	Muy Fina			
0,0625				
	Limo		Roca Lima	(>2/3 limo)
0,0039		Lodo	Lodo Calcáreo	(partes iguales de limo y arcilla)
	Arcilla		Roca Arcillosa	(>2/3 arcilla)

* milímetro

Prefixes for SI Unit Multiples
Prefijoas de Multiples de Unidades SI

exa (10^{18})	E
peta (10^{15})	P
tera (10^{12})	T
giga (10^{9})	G
mega (10^{6})	M
kilo (10^{3})	k
hecto (10^{2})	h
deka (10)	da
deci (10^{-1})	d
centi (10^{-2})	c
milli (10^{-3})	m
micro (10^{-6})	μ
nano (10^{-9})	n
pico (10^{-12})	p
femto (10^{-15})	f
atto (10^{-18})	a

STRATIGRAPHIC ABBREVIATIONS
ABREVACIÓNES STRATIGRÁFICOS

Term or lithology	Abbreviation	Termino o litología
Group	Gp.	Grupo
Formation	Fm.	Formación
Member	Mbr.	Miembro
Sandstone	Ss.	Arenisca
Siltstone	Slts.	Limolita
Shale	Sh.	Lutita
Limestone	Ls.	Caliza
Dolomite	Dol.	Dolomita
Conglomerate	Cgl.	Conglomerado
Breccia	Br.	Brecha
Quartzite	Qzt.	Cuarcita
Volcanics	Vol.	Rocas volcanicas
Claystone	Clyst.	Arcillita
Mudstone	Mdst.	Lutita
Granite	Gr.	Granito
Gneiss	Gn.	Gneis
Rhyolite	Rhy.	Riolita

Conversion Table of Weights and Measures
Tabla de Conversiones de Pesas y Medidas

inch (pulgada)	× 2.54	=	centimeter (centímetro)
centimeter (centímetro)	× 0.39	=	inch (pulgada)
inch (pulgada)	× 25.4	=	millimeter (milímetro)
millimeter (milímetro)	× 0.04	=	inch (pulgada)
foot (pie)	× 30.48	=	centimeter (centímetro)
centimeter (centímetro)	× 0.03	=	foot (pie)
foot (pie)	× 0.30	=	meter (metro)
meter (metro)	× 3.28	=	foot (pie)
sq. inch (pulgada cuadrada)	× 6.45	=	sq. centimeter (centímetro)
sq. centimeter (centímetro cuadrado)	× 0.16	=	sq. inch (pulgada cuadrada)
cubic inch (pulgada cúbica)	× 16.39	=	milliliters (mililitros)
milliliter (mililitros)	× 0.06	=	cubic inch (pulgada cúbica)
sq. foot (pie cuadrado)	× 0.09	=	sq. meter (metro cuadrado)
sq. meter (metro cuadrado)	× 10.76	=	sq. foot (pie cuadrado)
cubic foot (pulgada cúbica)	× 0.028	=	cubic meter (metro cúbico)
cubic meter (metro cúbico)	× 35.32	=	cubic foot (pie cúbico)

yard (yarda)	× 0.91	=	meter (metro)
meter (metro)	× 1.09	=	yard (yarda)
sq. yard (yarda cuadrada)	× 0.84	=	sq. meter (metro cuadrado)
sq. meter (metro cuadrado)	× 1.20	=	sq. yard (yarda cuadrada)
cubic yard (yarda cúbica)	× 0.76	=	cubic meter (metro cúbico)
cubic meter (metro cúbico)	× 1.31	=	cubic yard (yarda cúbico)
mile (milla)	× 1.61	=	kilometer (kilómetro)
kilometer (kilometro)	× 0.62	=	mile (milla)
sq. mile (milla cuadrada)	× 2.60	=	kilometer (kilómetro)
sq. kilometer (kilómetro cuadrado)	× 0.39	=	sq. mile (milla cuadrada)
acre (acre)	× 0.40	=	hectare (hectárea)
hectare (hectárea)	× 2.47	=	acre (acre)
ounce (onza)	× 28.35	=	gram (gramo)
gram (gramo)	× 0.035	=	ounce (onza)
pound (libra)	× 0.45	=	kilogram (kilogramo)
kilogram (kilogramo)	× 2.21	=	pound (libra)
ton (2,000 pounds) (tonelada)	× 0.91	=	metric ton (tonelada métrica)
metric ton (1,000 kg) (tonelada métrica)	× 1.10	=	ton (tonelada corta)
ounce (onza)	× 29.57	=	milliliter (mililitro)
milliliter (mililitro)	× 0.03	=	ounce (onza)
gallon (galón)	× 3.79	=	liter (litro)
liter (litro)	× 0.26	=	gallon (galón)
temperature *temperatura:* °F = 1.8 × °C + 32			°C = (°F − 32)1.8

References
Referencias

Bates, Robert L., and Julia A. Jackson, eds.
    1984. *Dictionary of Geological Terms*. 3d
    ed. New York: Doubleday.
*Collins Spanish-English, English-Spanish
    Dictionary*. 1994. New York: Harper-
    Collins.
Dictionary of Geology and Minerology. 1997.
    New York: McGraw-Hill.
Hansen, W. R., ed. 1991. *Suggestions to
    Authors of the Reports of the U.S. Geological
    Survey (STA)*. 7th ed. Washington, D.C.:
    USGS.
Nuevo Peqeño Larousse. 1951. Paris: Librairie
    Larousse.
Planet Earth. 1992. New York: Time Life.
Planeta Tierra. 1996. New York: Time Life
    Latinoamérica.

Prost, Gary L. 1997. *English-Spanish and Spanish-English Glossary of Geological Terms, Diccionario Inglés-Español y Español-Inglés de Términos de Geociencias.* Amsterdam: Overseas Publishers Association.

Turner, Juan Carlos M. 1972. *Diccionario Geológico, Inglés-Español, Español-Inglés.* Buenos Aires: Asociación Geológica Argentina.

Visser, W. A., ed. 1980. *Geological Nomanclature.* Haugue, Netherlands: Martinus Nijhoff.

Webster's New Universal Unabridged Dictionary, Second Edition. 1983. New York: Simon and Schuster.

Related Books from Sunbelt Publications

The Rise and Fall of San Diego: 150 Million Years of History Recorded in Sedimentary Rocks
By Patrick L. Abbott

Geology and Lore of the Northern Anza-Borrego Region: The Lows to Highs of Anza-Borrego Desert State Park
San Diego Association of Geologists
Edited by Monte L. Murbach and Charles E. Houser

Fossil Treasures of the Anza-Borrego Desert: The Last Seven Million Years
Edited by George T. Jefferson and Lowell Lindsay

The Sunrise Highway: A Geology Guide to San Diego's Laguna Mountains
San Diego Association of Geologists
By Michael J. Walawender

Geology of Anza-Borrego: Edge of Creation
By Paul Remeika and Lowell Lindsay

www.sunbeltbooks.com

2011